SpringerBriefs in Computer Science

Series Editors

Stan Zdonik
Peng Ning
Shashi Shekhar
Jonathan Katz
Xindong Wu
Lakhmi C Jain

For further volumes:
http://www.springer.com/series/10028

Adriano Veloso · Wagner Meira Jr.

Demand-Driven Associative Classification

Springer

Adriano Veloso
Computer Science Department
Universidade Federal de Minas Gerais
Antonio Carlos Av
Belo Horizonte
Brazil
e-mail: adrianov@dcc.ufmg.br

Wagner Meira Jr.
Computer Science Department
Universidade Federal de Minas Gerais
Antonio Carlos Av
6620 Belo Horizonte
Brazil
e-mail: meira@dcc.ufmg.br

ISSN 2191-5768

e-ISSN 2191-5776

ISBN 978-0-85729-524-8

e-ISBN 978-0-85729-525-5

DOI 10.1007/978-0-85729-525-5

Springer London Dordrecht Heidelberg New York

British Library Cataloguing in Publication Data
A catalogue record for this book is available from the British Library

Cover design: eStudio Calamar, Berlin/Figueres

Printed on acid-free paper

Springer is part of Springer Science+Business Media (www.springer.com)

Foreword

As president of the Brazilian Computer Society I am honored to introduce you all to *Demand-Driven Associative Classification*. The authors provide a fresh look over classification, and provide an innovative way in which to decompose hard problems into sub-problems and solve those individually. In the process, they run into several challenging questions, whose solution provide philosophical insights that should be of interest to all. The authors address how to define sub-problems, what particularities should be considered (if at all) and, my personal favorite, does independently solving sub-problems lead to better approximations than directly solving the entire problem?

The text provided here is an extension of the Ph.D. dissertation that won the first prize in the national CTD competition. The Thesis and Dissertation Competition is a national event, sponsored by the Brazilian Computer Society since 1987, whose goal is to award the very best Computer Science graduate work. Hundreds of submissions are received yearly, from over 50 different higher education institutions all over Brazil. Submissions are carefully reviewed by an international expert committee, to ensure that nothing but the highest quality standard work is selected for the national finals, that take place during the Society's annual Conference. The text that you are about to read underwent this rigorous process, and emerged as the winner.

We are also very happy with this partnership with Springer, whom we recognize for disseminating high quality computer science research worldwide.

São Carlos, December 2010 José Carlos Maldonado

Preface

The ultimate goal of computers is to help humans to solve problems. The solutions for such problems are typically programmed by experts, and the computers need only to follow the specified steps to solve the problem. However, the solution of some problems may be too difficult to be explicitly programmed. In such difficult cases, instead of directly programming the computers to solve the problem, they can be programmed to learn the solution. Machine Learning encompasses techniques used to program computers to learn. It is one of the fastest-growing research areas today, mainly motivated by the fact that the advent of improved learning techniques would open up many new uses for computers (i.e., problems for which the solution is hard to program by hand).

A prominent approach to machine learning is to repeatedly demonstrate how the problem is solved, and let the computer learn by example, so that it generalizes some rules about the solution and turn these into a program. This process is known as supervised learning. Specifically, the computer takes matched values of inputs (instantiations of the problem to be solved) and outputs (the solution) and models whatever information their relation contains in order to emulate the true mapping of inputs to outputs. When outputs are drawn from a pre-specified and finite set of possibilities, the process is known as classification. Under the assumption that inputs and outputs are related according to an unknown function, a classification problem may be seem as a function approximation problem: given some inputs for which the outputs (i.e., the classes) are known, the goal is to extrapolate the (unknown) outputs associated with other inputs as accurately as possible.

Several difficulties and impediments make classification problems challenging, motivating this work. The key insight that is discussed in this book is that a difficult classification problem can be decomposed into several, much simpler sub-problems. In this book we show that, instead of directly solving a difficult problem, independently solving its sub-problems by taking into account their particular demands, often leads to improved classification performance. This is shown empirically, by solving real-world problems (for which the solutions are hard to program) using the computationally efficient algorithms that are presented in this book. These problems include categorization of documents and name disambiguation in digital libraries, ranking documents retrieved by search engines, protein functional analysis, revenue optimization, among others. Improvements in classification performance are reported for all these problems (in some cases with gains of more than 100%). Further, theoretical evidence supporting our algorithms is also provided.

Belo Horizonte, December 2010   Adriano Veloso
                                                                      Wagner Meira Jr.

Acknowledgments

Many people deserve our thanks. We are extremely grateful to all our coauthors during the development of this work: Anderson Ferreira, Alberto Laender, Dorgival Guedes, Edleno de Moura, Eduardo Valle, Jussara Almeida, Marco Cristo, Marcos Gonçalves, Nivio Ziviani, Renato Ferreira, and Ricardo Torres.

We are also greatly indebted to Mohammed Zaki from the Rensselaer Polytechnic Institute for the pleasure of collaborating with him through these last years. Many thanks to José Carlos Maldonado, for writing the Foreword, and to the Brazilian Computer Society (SBC) for all the support towards publishing this book.

Special thanks go to the Computer Science Department at Universidade Federal deMinas Gerais (UFMG), for providing us with an ideal environment to work. Very special thanks go to UOL ("Universo Online"), for providing motivating application scenarios to us, in addition to the generous financial support through the "Bolsa UOL Pesquisa". Finally, we would like to thank CNPq, CAPES, Finep, and Fapemig for supporting partially this work.

Belo Horizonte, December 2010

Adriano Veloso
Wagner Meira Jr.

Contents

Part I
Introduction and Preliminaries

Chapter 1
Introduction

Abstract Learning is a fundamental ability of many living organisms. It leads to the development of new skills, values, understanding, and preferences. Improved learning capabilities catalyze the evolution and may distinguish entire species with respect to the activities they are able to perform. The importance of learning is, thus, beyond question. Learning covers a broad range of tasks. Some tasks are particularly interesting because they can be mathematically modeled. This makes natural to wonder whether computers might be made, or programmed, to learn (Turing, Can digital computers think? A talk on BBC Third Programme, 15 May 1951). A deep understanding of how to program computers to learn is still far, but it would be of great impact because it would increase the spectrum of problems that computers can solve. Candidate problems range between two extremes: structured problems for which the solution is totally defined [and thus are easily programmed by humans (Hutter, The fastest and shortest algorithm for all well-defined problems. Int J Found Comp Sci 13(3), 431–443 2002)], and random problems for which the solution is completely undefined, and thus cannot be programmed. Problems in the vast middle ground have solutions that cannot be well defined and are, thus, inherently hard to program (no one could foresee and provide solutions for all possible situations). Machine learning is one possible strategy to handle this vast middle ground and many tedious and difficult hand-coding tasks would be replaced by automatic learning methods. In this book we investigate an important learning method which is known as classification.

Keywords Machine learning · Classification · Mapping function · Function approximation · Problem decomposition · Association rules

A. Veloso and W. Meira Jr., *Demand-Driven Associative Classification*, SpringerBriefs in Computer Science, DOI: 10.1007/978-0-85729-525-5_1,
© Adriano Veloso 2011

1.1 Classification as a Function Approximation Problem

A prominent approach to machine learning is to provide to the computer examples demonstrating solutions for some situations or instantiations of the problem. These examples are paired values of inputs (instantiations of the problem) and outputs (the corresponding solution). Inputs and outputs are related somehow, but this relationship is unknown. The computer must generalize rules about this relationship and turn these rules into a program. This program will predict the outputs associated with inputs for which the solution is unknown. When the solution assumes pre-defined and finite values (which are called classes), this process is known as classification. Classification is a major task in predictive data mining [7]. According to [8], six out of the ten most influential data mining algorithms are classification algorithms.

The relationship between inputs and outputs may be expressed as a mapping function, which takes an input and provides the corresponding output. Since this function is unknown, the classification problem can be essentially stated as a function approximation problem: given as examples some inputs for which the outputs (i.e., the classes) are known, the goal is to extrapolate the (unknown) outputs associated with yet unseen inputs as accurately as possible. Several classification algorithms follow this function approximation paradigm [2, 4, 5]. These algorithms usually rely on a single mapping function to approximate the target function (i.e., the relationship between inputs and outputs). This single function is selected from a set of candidate functions and is the one which is most likely to provide the best available approximation to the target function. This implies that such single function will be used to approximate the target function over the full space of inputs. This is not necessarily a good strategy, because:

- the set of possible functions might not contain a good approximation of the target function for the full input space;
- the use of a single function to approximate the target function over the full space of inputs tends to be good on average, but it may fail on some particular regions of the input space.

Figure 1.1 illustrates the function approximation process. The left-most graph on the top shows the target function, where each point represents an input–output pair. The black points are given as examples to the classification algorithm, which uses them to build the mapping function. The white points are used to assess the accuracy of this function. Different mapping functions are shown. The right-most graph on the top shows a mapping function that does not provide a good approximation for the target function. Graphs on the bottom show mapping functions that fit well the target function, although they still fail on some particular regions of the input space.

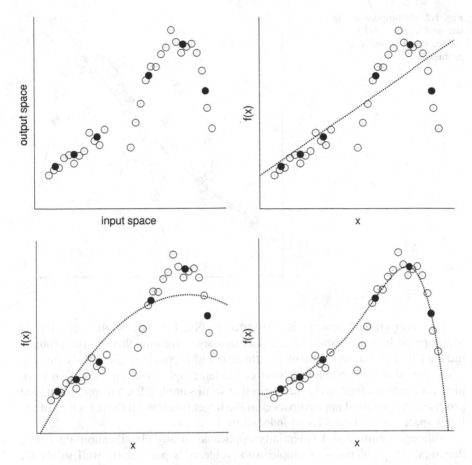

Fig. 1.1 An illustration of the classification problem

1.2 Problem Decomposition

The limiting factor of classification algorithms is the accuracy of the mapping functions they can provide in a reasonable time. Dramatic gains cannot be achieved through minor algorithmic modifications, but require the introduction of new strategies and approaches. The approach that we describe in this book consists in decomposing a hard classification problem into easier sub-problems. Each sub-problem is defined at test time, and it is induced by the input being classified. A specific mapping function is independently built for each sub-problem, on a demand-driven basis, by taking into account particularities of each sub-problem. This strategy leads to a finer-grained function approximation process, in which multiple mapping functions are built. Each mapping function is likely to perform particularly accurate predictions for the inputs that induce the corresponding sub-problems.

Fig. 1.2 Decomposition into
sub-problems. x_1 and x_2
induce two different sub-
problems

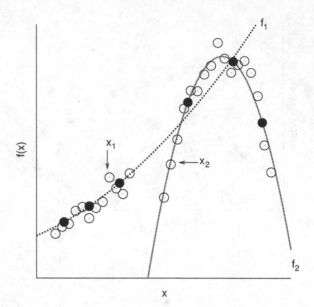

This finer-grained process is illustrated in Fig. 1.2. The original problem is decomposed into two sub-problems. Let us assume for now that the sub-problem induced by input x_1 consists of the three left-most samples (i.e., black points), while the other sub-problem consists of the three right-most samples. Two mapping functions are built using the respective set of samples. Each mapping function provides an optimized approximation of the target function on those regions of the input space (i.e., sub-problems) induced by x_1 and x_2.

Although intuitive and potentially applicable to any classification algorithm, decomposing problems into simpler sub-problems is particularly well suited for algorithms that follow the *associative classification* paradigm [3]. These algorithms look for *association* rules [1], that are local patterns and regularities hidden in the examples, and use these rules as building blocks to produce mapping functions. Specifically, given an input x_i, all rules carrying information related to x_i are combined in order to predict the correct output (i.e., class) for x_i. From the set of all possible rules that could be extracted from the given examples, only a small fraction will carry some information about a specific input x_i. The remaining rules are simply meaningless to x_i. This gives us a clue of how to induce a sub-problem for input x_i – the sub-problem induced by x_i is composed by examples from which we could extract only rules carrying information related to x_i. This is the key idea exploited by the algorithms that we describe in this book.

Intuitively, if x_i is an easy input, then it will induce a simple sub-problem, in the sense that we will be able to quickly find regularities that are strongly correlated to a specific output (i.e., most examples in the sub-problem will be associated with the same class). On the other hand, if x_i is a hard input, then it will induce a complex sub-problem, and we will have to spend more effort in order to find regularities that are strongly correlated with a specific output. Whatever the case, we discuss efficient algorithms, meaning that the number of rules grows polynomially with the of the sub-problem. Further, these algorithms are also [6], needing few examples to build accurate mapping functions. Much of this book is concerned with describing and evaluating these algorithms.

1.3 Overview of the Book

The structure of the book is as follows:

Chapter 2: We introduce some preliminaries on the problem. We review the classic theory on as a function approximation process, and we survey briefly the most important model selection procedures.

Chapter 3: We study the associative classification problem. We present simple associative classification algorithms, using the concepts presented in the previous chapter. We evaluate the proposed algorithms both empirically and theoretically.

Chapter 4: We propose our demand-driven associative algorithms. These algorithms break a difficult classification task into simpler sub-problems that can be solved independently. We evaluate the proposed algorithms both empirically and theoretically.

Chapter 5: We study the multi-label classification problem, and we propose multi-label demand-driven associative classification algorithms. The proposed algorithms employ the label propagation strategy which boosts classification performance.

Chapter 6: We evaluate different measures and metrics that capture the association between features and classes, and we introduce the metric domain of competence, which indicates the circumstances in which a metric is more appropriate that other available metrics. The best metric is selected using a stacking-like meta-learning procedure.

Chapter 7: We propose a strategy for calibrating the class membership probabilities provided by demand-driven associative classification algorithms. Our calibration mechanism is based on information gain and minimum description length.

Chapter 8: We investigate classification problems for which the acquisition of vast amounts of training examples is not feasible due to the cost associated with manual labeling process. We propose demand-driven associative classification algorithms that are able to incorporate new examples based on their own predictions.

Chapter 9: We investigate ordinal regression and ranking problems. Instead of predicting the correct class for the instances, our demand-driven associative classification algorithms are extended to sort instances according to the given criteria.

Chapter 10: We conclude the book and summarize our contributions. Further, we discuss opportunities for future work on associative classification.

References

1. Agrawal, R., Imielinski, T., Swami, A.: Mining association rules between sets of items in large databases. In: Proceedings of the International Conference on Management of Data (SIGMOD), pp. 207–216. ACM Press (1993)
2. Evgeniou, T., Pontil, M., Poggio, T.: Statistical learning theory: a primer. Int. J. Comp. Vis. **38**(1), 9–13 (2000)
3. Liu, B., Hsu, W., Ma, Y.: Integrating classification and association rule mining. In: Proceedings of the Conference on Data Mining and Knowledge Discovery (KDD), pp. 80–86. ACM Press (1998)
4. Poggio, T., Girosi, F.: A sparse representation for function approximation. Neural Comput. **10**(6), 1445–1454 (1998)
5. Rahimi, A., Recht, B.: Uniform approximating functions with random bases. In: Allerton, pp. 43–49. SIAM (2008)
6. Valiant, L.: A theory of the learnable. Commun. ACM **27**(11), 1134–1142 (1984)
7. Witten, I., Frank, E.: Data mining: practical machine learning tools and techniques. Morgan Kaufmann, San Francisco (2005)
8. Wu, X., Kumar, V., Quinlan, J., Ghosh, J., Yang, Q., Motoda, H., McLachlan, G., Ng, A., Liu, B., Yu, P., Zhou, Z., Steinbach, M., Hand, D., Steinberg, D.: Top 10 algorithms in data mining. Knowl. Inf. Syst. **14**(1), 1–37 (2008)

Chapter 2
The Classification Problem

Abstract We present the classification problem, starting with definitions and notations that are necessary to ground posterior discussions. Then, we discuss the *Probably Approximately Correct* learning framework, and some function approximation strategies.

Keywords Generalization · PAC-learning · Sample complexity · Classification efficiency · Empirical risk minimization · Structural risk minimization

2.1 Definitions

In this section we present definitions and notations that form the basis of the classification problem.

2.1.1 Training Data and Test Set

In a classification problem, there is a set of input-output pairs (also referred to as instances or examples) of the form $z_i = (x_i, y_i)$. Each input x_i is a fixed-length record of the form $\langle a_1, \ldots, a_l \rangle$, where each a_i is an attribute-value. Each output y_i draws its value from a discrete and finite set of possibilities $y = \{c_1, \ldots, c_p\}$, and indicates the class to which z_i belongs. Cases where $y_i = ?$ indicate that the correct class of z_i is unknown. There is a fixed but unknown conditional probability distribution $P(y|x)$, that is, the relationship between inputs and outputs is fixed but unknown. The set of pairs is explicitly divided into two partitions, the training data (denoted as S) and the test set (denoted as T):

A. Veloso and W. Meira Jr., *Demand-Driven Associative Classification*,
SpringerBriefs in Computer Science, DOI: 10.1007/978-0-85729-525-5_2,
© Adriano Veloso 2011

Fig. 2.1 Simple and complex mapping functions

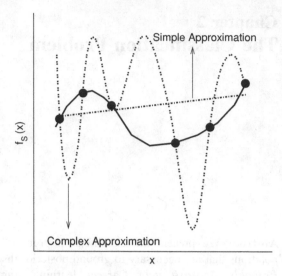

$$S = \{s_1 = (x_1, y_1), \ldots, s_n = (x_n, y_n)\}$$
$$T = \{t_1 = (x_{n+1}, ?), \ldots, t_m = (x_{n+m}, ?)\}$$

Further, it is assumed that pairs in T are in some sense related to pairs in S, and that $\{t_{n+1}, \ldots, t_{n+m}\}$ and $\{s_1, \ldots, s_n\}$ are sampled independently and identically from the same distribution $P(y|x)$.

2.1.2 Classification Algorithm

A classification algorithm takes as input the training data S and the test set T, and returns a mapping function $f_S : x \rightarrow y$ that represents the relation between inputs and outputs in S, that is, the mapping function f_S is a discrete approximation of $P(y|x)$ (i.e., a classification algorithm observes n input–output pairs and produces a function which describes well the underlying input–output process). Many possible functions may be derived from S. The hypothesis space H is the space of functions explored by the classification algorithm in order to select f_S. The selected mapping function f_S is finally used to estimate the outputs y given the inputs x, for each $x_i \in T$.

Figure 2.1 illustrates the problem of function approximation. The dark solid line represents the true (target) function. The dark points are given as examples (i.e., $S = \{s_1, \ldots, s_n\}$). Two approximations (i.e., candidate functions) are used to fit the true function. The complex approximation fits S exactly. Yet, it is clear that the complex approximation will perform poorly in T, as it is far from the true function on most of the space of inputs (i.e., the x-axis). The simple approximation does not fit S exactly, but provides better approximations for most of the points in T. The classification problem is that of selecting, from all functions in H, the one which best approximates (discretely) the distribution $P(y|x)$. The selection is based on S.

This formulation implies that the classification problem corresponds to the problem of function approximation.

2.1.3 Loss Function

A loss function, $\ell(f_S, z_i = (x_i, y_i))$, represents the loss (or cost) associated with a wrong estimate (i.e., $f_S(x_i) \neq y_i$) as a function of the degree of deviation from the correct value. Unless otherwise stated, the 0–1 loss function will be the one used throughout this thesis, where for $z_i = (x_i, y_i)$:

$$\ell(f_S, z_i) = \begin{cases} 0 & \text{if } f_S(x_i) = y_i \\ 1 & \text{otherwise} \end{cases}$$

The 0–1 loss function is very intuitive, since it states that one should make as few mistakes as possible. It may be considered an upper bound for other loss functions, such as the hinge and the squared loss functions [11].

2.1.4 Expected Error

According to [4] and [15], the expected error of a mapping function f_S is defined as:

$$I_T[f_S] = \int_{t=(x,y)} \ell(f_S, t) \mathrm{d}P(y|x)$$

The primary goal of classification algorithms is to select a mapping function f_S for which $I_T[f_S]$ is guaranteed low. However, $I_T[f_S]$ cannot be computed because the conditional probability distribution $P(y|x)$ is unknown.

2.1.5 Empirical Error

Although the expected error is unknown, the empirical error of a mapping function f_S can be easily computed using S:

$$I_S[f_S] = \frac{1}{n} \sum_{i=1}^{n} \ell(f_S, s_i)$$

2.1.6 Generalization

An important ability for any classification algorithm is generalization: the empirical error must converge to the expected error as the number of examples n

increases, that is, $I_S[f_S] \approx I_T[f_S]$. Informally, the classification performance of the selected function, f_S, in S must be a good indicator of its classification performance in T. Generalization error (or risk), denoted as ε, is given by $I_T[f_S] - I_S[f_S]$. High generalization (i.e., low values of ε) implies low expected error only if $I_S[f_S] \approx 0$.

Next we discuss a well-known mathematical tool for the analysis of classification algorithms.

2.2 Probably-Approximately Correct (PAC) Learning

The Probably-Approximately Correct (PAC) learning framework [12, 13] states that the classification algorithm must be able to select a mapping function f_S from H which, with high probability, will have low expected error. There are two major requirements in the PAC learning framework:

- The expected error is bounded by some constant ε (i.e., the generalization error).
- The probability that the expected error is greater than ε is bounded by some constant δ.

Putting simple, the PAC learning framework requires that the classification algorithm *probably* selects a mapping function f_S that is *approximately correct*. More specifically, a classification problem is PAC-feasible if the algorithm selects a mapping function $f_S \in H$, such that $I_T[f_S] \leq \varepsilon$, with probability of at least $(1 - \delta)$, for $0 < \varepsilon < \frac{1}{2}$ and $0 < \delta < \frac{1}{2}$. This statement is formalized as follows:

$$P[I_T[f_S] < \varepsilon] \geq 1 - \delta \qquad (2.1)$$

2.2.1 Sample Complexity

The sample complexity of a classification algorithm is the relation between $I_T[f_S]$ and $|S|$ (or n). Inequality (2.1) can be used to derive the sample complexity of a classification algorithm. In this case, a mapping function, f_S, is considered accurate if $I_T[f_S] < \varepsilon$. We denote an accurate function as f^+, and similarly, we denote poor functions as f^-. Also, f^* is the most accurate mapping function in the hypothesis space H.

For a given pair $z_i = (x_i, y_i)$ (i.e., an example), the probability of $f^-(x_i) \neq y_i$, is at least ε. Thus, the probability of $f^-(x_i) = y_i$ is at most $1 - \varepsilon$. So, for n pairs $\{z_1 = (x_1, y_1), \ldots, z_n = (x_n, y_n)\}$, the probability that $f^-(x_1) = y_1 \wedge \cdots \wedge f^-(x_n) = y_n$ is at most $(1 - \varepsilon)^n$. Now, considering that there are k poor functions in H, the probability that at least one of these functions correctly predicts the output of the n pairs is $k \times (1 - \varepsilon)^n$. Using the fact that $k \leq |H|$ (and assuming that $I_T[f^*] = 0$), the following inequality is obtained:

$$P[I_T[f_S] > \varepsilon] \leq H \times (1 - \varepsilon)^n \leq \delta \qquad (2.2)$$

Since $(1 - \varepsilon) \leq e^{-\varepsilon}$ [7], and solving for n, (2.2) can be rewritten as:

$$P[I_T[f_S] > \varepsilon] \leq H \times e^{-n\varepsilon} \leq \delta$$
$$H \times e^{-n\varepsilon} \leq \delta$$
$$n \geq \frac{1}{\varepsilon}\left(\ln|H| + \ln\left(\frac{1}{\delta}\right)\right) \qquad (2.3)$$

Thus, the more accuracy (lower ε values) and the more certainty (lower δ values) one wants, the more examples the classification algorithm needs. Now, (2.2) and (2.3) can be used to derive the expected error bound:

$$\varepsilon \geq \frac{1}{n}\left(\ln|H| + \ln\left(\frac{1}{\delta}\right)\right)$$
$$I_T[f_S] \leq I_S[f_S] + \frac{1}{n}\left(\ln|H| + \ln\left(\frac{1}{\delta}\right)\right) \qquad (2.4)$$

So far, it was assumed that $I_S[f^*] = 0$ (i.e., the classification algorithm is gnostic[1]). If $I_S[f^*] > 0$ (i.e., the classification algorithm is agnostic), then, according to [1], the Chernoff approximation can be used to derive the sample complexity:

$$n \geq \frac{1}{2\varepsilon^2}\left(\ln|H| + \ln\left(\frac{1}{\delta}\right)\right) \qquad (2.5)$$

Now, (2.2) and (2.5) can be used to derive the expected error bound:

$$\varepsilon \geq \sqrt{\frac{1}{2n}\left(\ln|H| + \ln\left(\frac{1}{\delta}\right)\right)}$$
$$I_T[f_S] \leq I_S[f_S] + \sqrt{\frac{1}{2n}\left(\ln|H| + \ln\left(\frac{1}{\delta}\right)\right)} \qquad (2.6)$$

For PAC-based expected error bounds, $|H|$ must be estimated. The simpler the hypothesis space (or, equivalently, the fewer functions are explored), the lower is ε, at the expense of increasing the empirical error.

2.2.2 Classification Efficiency

The empirical error is a finite sample approximation of the expected error. It can be shown [4] that the empirical error converges uniformly to the expected error when $|S| \to \infty$ ($n \to \infty$). An efficient classification algorithm ensures that this

---

[1] A function f_S is consistent with example $s = (x, y)$ if $f_S(x) = y$. A classification algorithm is gnostic if it selects a function f_S which is consistent with all examples in S.

convergence occurs with high rate. Formally, in the PAC learning framework, a classification algorithm is efficient if it selects, in polynomial time and with a polynomial number of examples, with probability $(1 - \delta)$, a function $f_S \in H$ for which $I_S[f_S] < \varepsilon$, and $I_S[f_S] \approx I_T[f_S]$ (that is, efficient classification algorithms must achieve low empirical error, with access to a restricted number of examples and in a reasonable amount of time).

2.3 Function Approximation

Classification is posed as synthesizing a mapping function that best approximates the relationship between the inputs x_i and the corresponding outputs y_i (i.e., the classes). Two strategies for function approximation are considered in this work: empirical risk minimization (which follows the stability theory [3, 5, 8, 9]), and structural risk minimization (which follows the VC theory [6, 14, 15]). Both strategies establish sufficient conditions for generalization. Next we will discuss these strategies.

2.3.1 Empirical Risk Minimization

Probably the most natural function approximation strategy is Empirical Risk Minimization (ERM): from all possible mapping functions in H, the classification algorithm selects the function f_S that minimizes $I_S[f_S]$, the empirical error given by:

$$\arg \min \left(\frac{1}{n} \sum_{i=1}^{n} \ell(f_S, s_i) \right), \quad \forall f_S \in H \tag{2.7}$$

The Empirical Risk Minimization strategy, however, does not ensure generalization. More specifically, minimizing the empirical error does not necessarily imply in minimizing the expected error. A sufficient condition for generalization of ERM algorithms is the stability of f_S [9, 10].

Stability—The stability measures the difference, β_{s_i}, in empirical errors at a pair $s_i \in S$ between a function f_S obtained given the entire training data S and a function f_{S-s_i} obtained given the same training data but with pair s_i left out.

Specifically, if the training data S is perturbed by removing one pair s_i, and if the selected function f_S does not diverge much from f_{S-s_i}, then f_S is stable. Informally, avoiding unstable functions can be thought as a way of controlling the variance of the function approximation process. Function f_S is β-stable if:

$$\forall s_i \in S, |f_S(s_i) - f_{S-s_i}(s_i)| \leq \beta \tag{2.8}$$

The lowest value of β in (2.8) provides the stability of f_S. The lowest value of β is the largest change at any pair s_i. Thus, function f_S shown in Fig. 2.2, is obtained by Empirical Risk Minimization using $S = \{s_1, s_2, s_3, s_4, s_5\}$. Similarly, function

Fig. 2.2 Empirical risk
minimization using stability

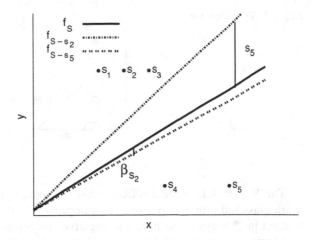

f_{S-s_2} is obtained by Empirical Risk Minimization using $\{S - s_2\}$, and function f_{S-s_5} is obtained by Empirical Risk Minimization using $\{S - s_5\}$. The difference at s_2, β_{s_2} is small. The difference at s_5, β_{s_5}, is large. Therefore, f_S is β_{s_5}-stable, despite the very small value of β_{s_2}. Function f_S is stable if $\beta = O(\frac{1}{n})$.

It has been shown in [3] that the expected error can be estimated by the empirical error and the stability of the selected function f_S, as follows:

$$I_T[f_S] \leq I_S[f_S] + \left(\beta + (4n\beta + 1) \times \sqrt{\frac{\ln \frac{1}{\delta}}{2n}} \right) \qquad (2.9)$$

Thus, the function f_S that minimizes $I_T[f_S]$ may be selected by applying (2.9) to each possible candidate function.

2.3.2 Structural Risk Minimization

Structural Risk Minimization (SRM) provides a trade-off between the complexity of a function and its empirical error. Simpler functions may provide high empirical error (they may underfit the training data), while complex functions may provide low empirical error (but it may be by means of overfitting the training data). Thus, from all possible functions in H, the classification algorithm following the Structural Risk Minimization strategy selects the function f_S that minimizes this trade-off.

A structure is a (possibly infinite) nested set of classes of functions F_i, such that $F_1 \subseteq F_2 \subseteq \cdots$, where functions in F_1 are simpler (i.e., have lower complexity) than functions in $F_2 - F_1$, and so on. Since classes of functions are nested, the empirical error decreases as the complexity of classes of functions increases.

Fig. 2.3 VC-dimension

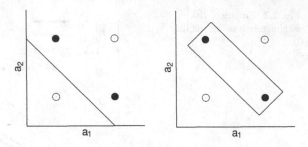

The Vapnik–Chervonenkis Dimension − Suppose $y_i \in \{\pm 1\}$ (i.e., examples in S are classified either as positive or as negative). In this case, n pairs in S can be labeled in 2^n ways as positive and negative. Therefore, 2^n different classification problems can be defined by n pairs. If for any of these problems, the function $f_S \in H$ exactly separates all the positive examples from all the negative ones, then it is said that f_S shatters n pairs (i.e., all pairs in f_S can be classified with no error by f_S). The maximum number of pairs in S that can be shattered by f_S is called the VC-dimension [2, 16] of f_S, which is denoted as d_{f_S}.

An example of classification problem composed of four pairs in two dimensions (i.e., attributes a_1 and a_2) is given in Fig. 2.3. A rectangle can shatter four points in two dimensions, but a line can shatter only three. Thus, for this classification problem, the VC-dimension of a rectangle is four, while the VC-dimension of a line is three. For a given classification problem, d_{f_S} is given by the maximum number of pairs that can be correctly classified by f_S. VC dimension may seem pessimistic, since it establishes that a line can only classify problems composed of three pairs, and not more. A function that can classify only three pairs is not very useful. However, this is because the VC dimension is independent of the probability distribution from which pairs are drawn (i.e., $P(y|x)$). In practice, however, pairs that are close to each other often have the same label.

Complexity − The VC-dimension measures the expressive power, richness or flexibility of a set of functions by assessing how wiggly its members can be. It has been shown [6] that the expected error can be estimated by the empirical error and the complexity of the selected function f_S, as follows:

$$I_T[f_S] \le I_S[f_S] + \sqrt{\frac{d_{f_S}\left(\ln\frac{2n}{d_{f_S}} + 1\right) - \ln\frac{\delta}{4}}{n}} \tag{2.10}$$

The shape of this bound, which is shown in Fig. 2.4, can be exploited to select a function with the most appropriate complexity for S. Classes of functions are considered in increasing complexity (i.e., F_1 is considered before F_2, and so on). Structural Risk Minimization corresponds to finding the simplest function f_S which provides the lowest empirical error. By applying (2.10) to each class of functions, a function in the class for which the error bound is tightest can be selected.

Fig. 2.4 Structural risk minimization using VC-dimension

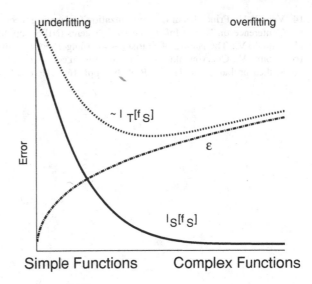

References

1. Angluin, D.: Computational learning theory: survey and selected bibliography. In: Proceedings of the Annual Symposium on Theory of Computing (STOC), pp. 351–369. ACM Press (1992)
2. Blumer, A., Ehrenfeucht, A., Haussler, D., Warmuth, M.: Learnability and the Vapnik–Chervonenkis dimension. Commun. ACM **36**(4), 865–929 (1989)
3. Bousquet, O., Elisseeff, A.: Stability and generalization. J. Mach. Learn. Res. **2**, 499–526 (2002)
4. Cucker, F., Smale, S.: On the mathematical foundations of learning. Bull. (New Series) Amer. Math. Soc. **39**(1), 1–49 (2001)
5. Devroye, L., Wagner, T.: Distribution-free performance bounds for potential function rules. Trans. Inf. Theory **25**(5), 601–604 (1979)
6. Guyon, I., Boser, B., Vapnik, V.: Automatic capacity tuning of very large vc-dimension classifiers. In: Proceedings of the Annual Conference on Neural Inf. Processing Systems (NIPS), pp. 147–155. MIT Press, Cambridge (1992)
7. Kearns, M., Vazirani, U.: An Introduction to Computational Learning Theory. MIT Press, Cambridge (1994)
8. Kutin, S., Niyogi, P.: Almost-everywhere algorithmic stability and generalization error. In: Proceedings of the Conference on Uncertainty in Artificial Intelligence (UAI), pp. 275–282 (2002)
9. Mukherjee, S., Niyogi, P., Poggio, T., Rifkin, R.: Learning theory: stability is sufficient for generalization and necessary and sufficient for consistency of empirical risk minimization. Adv. Comput. Math. **25**(1–3), 161–193 (2006)
10. Poggio, T., Rifkin, R., Mukherjee, S., Niyogi, P.: General conditions for predictivity in learning theory. Nature **428**, 419–422 (2004)
11. Rosasco, L., Vito, E.D., Caponnetto, A., Piana, M., Verri, A.: Are loss functions all the same? Neural Comput. **16**(5):1063–107 (2004)
12. Valiant, L.: A theory of the learnable. Commun. ACM **27**(11), 1134–1142 (1984)
13. Valiant, L.: A theory of the learnable. In: Proceedings of the Annual Symposium on Theory of Computing (STOC), pp. 436–445. ACM Press (1984)

14. Vapnik, V.: Principles of risk minimization for learning theory. In: Proceedings of the Annual Conference on Neural Inf. Processing Systems (NIPS), pp. 831–838. MIT Press (1991)
15. Vapnik, V.: The Nature of Statistical Learning Theory. Springer, Heidelberg (1995)
16. Vapnik, V., Chervonenkis, A.: On the uniform convergence of relative frequencies of events to their probabilities. Theory Probab. Appl. **16**(2), 264–280 (1971)

Part II
Associative Classification

Chapter 3
Associative Classification

Abstract The hypothesis space, H, may contain a huge (possibly infinite) number
of functions. Randomly producing functions, in the hope of finding one that
approximates well the target function $P(y|x)$, is not likely to be an efficient
strategy. Fortunately, there are countless more efficient strategies for producing
approximations of $P(y|x)$. One of these strategies is to directly exploit relation-
ships, dependencies and associations between inputs and outputs (i.e., classes) (Liu
et al. Integrating classification and association rule mining. In: Proceedings of the
Conference on Data Mining and Knowledge Discovery (KDD), 1998). Such
associations are usually hidden in the examples in S, and, when uncovered, they
may reveal important aspects concerning the underlying phenomenon that gen-
erated these examples (i.e., $P(y|x)$). These aspects can be exploited for the sake of
producing only functions that provide potentially good approximations of $P(y|x)$
(Cheng et al. Discriminative frequent pattern analysis for effective classification.
In: Proceedings of the International Conference on Data Engineering (ICDE),
2007; Cheng et al. Direct discriminative pattern mining for effective classification.
In: Proceedings of the International Conference on Data Engineering (ICDE),
2008; Fan et al. Direct mining of discriminative and essential frequent patterns via
model-based search tree. In: Proceedings of the Conference on Data Mining and
Knowledge Discovery (KDD), 2008; Li et al. Efficient classification based on
multiple class-association rules. In: Proceedings of the International Conference on
Data Mining (ICDM), 2001). This strategy has led to a new family of classification
algorithms which are often referred to as associative classification algorithms. The
mapping functions produced by these algorithms are composed of rules $X \rightarrow c_j$,
indicating an association between X, which is a set of attribute-values, and a class
$c_j \in y$. In this chapter, we present and evaluate novel associative classification
algorithms.

Keywords Hypothesis space · Discretization · Decision rules · Eager rule
extraction · Stability · VC-dimension · Document categorization

A. Veloso and W. Meira Jr., *Demand-Driven Associative Classification*,
SpringerBriefs in Computer Science, DOI: 10.1007/978-0-85729-525-5_3,
© Adriano Veloso 2011

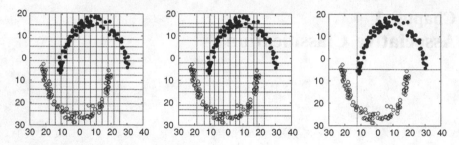

Fig. 3.1 Discretized input space

3.1 Continuous and Discrete Data

Attributes can assume continuous or discrete values. Associative classification algorithms require that continuous attributes are first discretized. Discretization refers to the process of splitting the space of attribute-values into intervals, where each interval contains values that carry (almost) the same information. The crucial point during discretization is the definition of the boundaries of each interval. Next we discuss three discretization methods.

3.1.1 Uniform Range and Uniform Frequency Discretization

A simple discretization method is to split the space of values of each attribute into equal, predefined interval ranges. This method is known as Uniform Range Discretization. Figure 3.1(left) shows the intervals defined by this method. Another simple strategy is to split the space of values of each attribute into variable-size intervals, so that each interval contains approximately the same number of examples. This method is known as Uniform Frequency Discretization. Figure 3.1(middle) shows the intervals defined by this method.

Since these methods do not use information about the input–output relationship in order to set interval boundaries, it is likely that important information will be lost as a result of combining values that are strongly associated with different classes into the same interval. This may be harmful to classification effectiveness.

3.1.2 Discretization Based on Minimum Description Length

A more sophisticated discretization method was proposed by [7]. It recursively splits the space of values of each attribute. The boundaries are those that provide the maximum information gain, that is, the intervals are found in a way that minimizes the entropy of the classes (e.g., intervals tend to include examples of the prevailing class). Splitting continues until the gain is below the minimal description length [15] of the interval. This may result in an arbitrary number of intervals, including a single interval in which case the attribute is discarded as useless. Figure 3.1 (Right) shows the intervals defined by this method.

3.2 Association Rules and Decision Rules

Association rules are implications of the form $a \rightarrow b$, where a and b are conjunctions of features/classes (i.e., any non-empty combination of features or classes). A much smaller subset of all association rules are particularly interesting for the sake of approximating $P(y|x)$. This subset is composed of rules, which hereafter will be called *decision rules*. Decision rules are implications of the form $X \rightarrow c_j$, where X is any combination of features, and $c_j \in y$ is a class. Thus, these implications can be considered local mappings from inputs to outputs (i.e., local patterns).

Decision rules are extracted from S. Thus, a decision rule $X \rightarrow c_j$ only exists if the set of features X is a subset of at least one input $x_i \in S$. Next we discuss some important properties of decision rules:

Confidence: A decision rule $X \rightarrow c_j$ is only interesting for the sake of approximating $P(y|x)$ if X and c_j are somehow associated. The confidence of a decision rule $X \rightarrow c_j$, which is denoted as $\theta(X \rightarrow c_j)$, is an indication of how strongly X and c_j are associated. It is given by the fraction of inputs containing X as subset, for which the corresponding output is c_j. Formally:

$$\theta(X \rightarrow c_j) = \frac{|(x_i, y_i)| \in S \quad \text{such that } X \subseteq x_i \text{ and } c_j = y_i}{|(x_i, y_i)| \in S \quad \text{such that } X \subseteq x_i} \quad (3.1)$$

Support: The support of a decision rule $X \rightarrow c_j$, which is denoted as $\sigma(X \rightarrow c_j)$, is an important indication of the reliability of the association between X and c_j. It is given by the fraction of inputs in S containing X as a subset, and having c_j as the corresponding output. Formally:

$$\sigma(X \rightarrow c_j) = \frac{|(x_i, y_i)| \in S \text{ such that } X \subseteq x_i \text{ and } c_j = y_i}{n} \quad (3.2)$$

Rule Complexity: There is no universal way of measuring complexity, and the choice of a specific measure for complexity is inherently dependent on the scenario of interest. Often, however, the length of an explanation can be an indication of its complexity [5, 10, 15, 16]. A decision rule $X \rightarrow c_j$ can be viewed as an explanation that leads to the decision (or a prediction) c_j. In this case, the explanation is essentially the features in X. Feature sets composed of few features provide simple explanations about the decision, while more complex explanations require more features. The size of the decision rule is the number of features in X (i.e., $|X|$) and is regarded as the complexity of the corresponding explanation.

Rule Usefulness: A decision rule $X \rightarrow c_j$ matches an input $x_i \in T$ if and only if $X \subseteq x_i$. Since the prediction provided by a decision rule is based on all the features that are in X, only decision rules matching input x_i are used to estimate the corresponding output y_i. A decision rule is useful for sake of approximating $P(y|x)$ if it matches at least one input in T, otherwise the decision rule is said to be useless.

3.3 Method and Algorithms

In this section a simple method for associative classification will be presented. Similar methods were already proposed in [12, 13]. First, it will be described how decision rules are extracted from S, and then how these decision rules are used in order to approximate $P(y|x)$. The method to be presented will hereafter be referred to as EAC (standing for Eager Associative Classification). Algorithms based on the EAC method are also presented in this section.

3.3.1 Level-Wise Rule Extraction

Simpler decision rules are extracted before more complex ones. Specifically, decision rules of size k are extracted from S before decision rules of size $k + 1$. The set of rules of size k is denoted as F_k. The support of a decision rule $X \rightarrow c_j$ is calculated by directly accessing S and counting the number of inputs having X as subset, for which c_j is the corresponding output.

3.3.1.1 Support Counting

Efficient support counting is mandatory for the scalability of associative classification algorithms. The main objective is to reduce the number of accesses to S. The algorithms we will present in this section employ an efficient strategy for support counting, which is based on vertical data representation and fast operations.

A vertical bitmap is created for each feature and class (or output) in S. That is, S is translated into a bitmap matrix, in which the columns are indexed by features (or classes), and lines are indexed by examples. If feature (or class) a_i appears in example s_j, then the bit indexed by example s_j and feature (or class) a_i is set to one; otherwise, the bit is set to zero.

Given the bitmap for feature a_i, and the bitmap for feature a_j, the bitmap for feature-set $\{a_i, a_j\}$ is simply the bitwise AND of these two bitmaps. Classes are processed in the same way. Each bitmap is divided into words, and each word is 16-bit length (i.e., each word comprises 16 examples).

All possible results of a bit-wise operation are stored in a look-up table with 2^{16} entries. After a word is processed, the corresponding result for this word is retrieved from the look-up table. The results associated with each word are finally summed. The final sum is the support count.

3.3.1.2 Pruning

The number of all possible decision rules hidden in S may be huge. Very often, practical limitations prevent the extraction of all such decision rules. Thus, pruning strategies must be employed in order to reduce the number of decision rules that are processed. EAC-based algorithms employ the typical pruning strategy which is to use a single cut-off value based on a minimum support threshold, σ_{min}. This cut-off value separates frequent from non-frequent decision rules. Specifically, only decision rules occurring at least $\sigma_{min} \times |S|$ times in S, are considered frequent. Since decision rules are extracted in a level-wise, bottom-up way, only feature-sets associated with frequent rules of size k are used to produce rules of size $k + 1$ [1] (i.e., only feature-sets included in rules in F_k are used to produce rules in F_{k+1}).

3.3.2 Prediction

Mapping functions perform predictions using the decision rules extracted from S, as it will be discussed next.

3.3.2.1 Grouping

After extracting decision rules from S, EAC-based algorithms group them in order to facilitate fast access to specific ones. In the discussion that follows it will be denoted as R an arbitrary set of decision rules extracted from S (i.e., frequent rules in S). Similarly, it will be denoted as R^{x_i} the subset of R which contains only the decision rules matching an arbitrary input x_i. Finally, it will be denoted as $R_{c_j}^{x_i}$ the subset of R^{x_i} which contains only decision rules predicting class c_j (i.e., decision rules of the form $X \rightarrow c_j$ for which $X \subseteq x_i$).

3.3.2.2 Learning Mapping Functions

Given an arbitrary input x_i, $f_S(x_i)$ gives the predicted output for x_i. The value of $f_S(x_i)$ is calculated based on the decision rules in R^{x_i}. Different algorithms based on the EAC method can be distinguished depending on the prediction strategy used.

Algorithm EAC-SR This algorithm (standing for EAC using a single rule) produces a mapping function, f_S, using frequent decision rules in S. Then, given an input x_i, f_S simply returns the output (or the class) predicted by the strongest decision rule (i.e., the one with highest confidence value) matching x_i, that is:

$$f_S(x_i) = c_j \text{ such that } \theta(X \to c_j) \text{ is argmax}(\theta(r)) \ \forall r \in R^{x_i} \tag{3.3}$$

EAC-SR is a PAC-efficient classification algorithm, as will be demonstrated next.

Theorem 3.1. *EAC-SR is PAC-efficient.*

Proof. Let q be the number of features in S. In this case, the hypothesis space H for EAC-SR is the set of all possible decision rules in S, which is clearly $|H| = p \times 2^q$. According to Eq. 2.4

$$n \geq \frac{1}{\epsilon}\left(\ln(p \times 2^q) + \ln\left(\frac{1}{\delta}\right)\right)$$

$$\geq \frac{1}{\epsilon}\left(\ln(p) + 0.69q + \ln\left(\frac{1}{\delta}\right)\right) \qquad \square$$

Thus, the sample complexity increases only polynomially with q. It is also necessary to show that a function f_S, for which $I_S[f_S] = 0$ is found in time polynomial in q.

Obviously, the number of functions in S is exponential in q (i.e., $O(2^q)$). However, since an input is composed of l attribute-values (or features), the maximum allowed size of decision rules can be held fixed to l (i.e., a decision rule can have at most l features). Thus, the number of possible decision rules in S is $p \times \left(q + \binom{q}{2} + \cdots + \binom{q}{l}\right) = O(q^l)$, and, thus, f_S can be found in time polynomial in q, according to Algorithm 1. This completes the proof.

Algorithm 1 Finding f_S, according to EAC-SR.

Require: The training data S, σ_{min}, and l
Ensure: f_S, if it exists. Failure, otherwise.

1: $R \Leftarrow$ rules $\{X \to c_j\}$ extracted from S, such that $|X| \leq l$ and $\sigma(X \to c_j) \geq \sigma_{min}$
2: **if** $\forall x_i \in S \ \exists r \in R^{x_i}$ such that $\theta(r) = 1.00$ **then**
3:     return f_S such that $f_S(x_i) = c_j$, where $\{X \to c_j\} \in R^{x_i}$ and $\theta(X \to c_j) = 1.00$
4: **else**
5:     return failure.
6: **end if**

An immediate question is to find under which circumstances $I_S[f_S] = 0$.

Theorem 3.2. *If $\sigma_{min} \leq \frac{1}{|S|}$, then $I_S[f_S] = 0$.*

Proof. If $\sigma_{min} \leq \frac{1}{|S|}$ then any feature set occurring at least once in S is frequent. Since S is composed of distinct inputs (i.e., instances with different features sets),

Table 3.1 Training data and test set given as example

		Input (x_i)			Output (y_i)
		a_1	a_2	a_3	
	(x_1, y_1)	[0.00–0.22]	[0.33–0.71]	[0.00–0.35]	1
	(x_2, y_2)	[0.00–0.22]	[0.33–0.71]	[0.35–1.00]	0
	(x_3, y_3)	[0.46–1.00]	[0.71–1.00]	[0.35–1.00]	1
	(x_4, y_4)	[0.22–0.46]	[0.00–0.33]	[0.35–1.00]	0
S	(x_5, y_5)	[0.00–0.22]	[0.33–0.71]	[0.00–0.35]	1
	(x_6, y_6)	[0.22–0.46]	[0.33–0.71]	[0.00–0.35]	1
	(x_7, y_7)	[0.00–0.22]	[0.33–0.71]	[0.00–0.35]	1
	(x_8, y_8)	[0.22–0.46]	[0.71–1.00]	[0.00–0.35]	1
	(x_9, y_9)	[0.46–1.00]	[0.00–0.33]	[0.35–1.00]	1
	(x_{10}, y_{10})	[0.22–0.46]	[0.00–0.33]	[0.35–1.00]	0
T	(x_{11}, y_{11})	[0.22–0.46]	[0.33–0.71]	[0.35–1.00]	?(1)

there must be feature sets occurring only once in S. A feature set, X, occurring once in S (an extreme case is $X = x_i$) must produce a decision rule $X \to c_j$ such that $\theta(X \to c_j) = 1.00$. Further, for any feature set Y such that $Y \subseteq X$, if there is a decision rule $Y \to \overline{c_j}$ then $\theta(Y \to \overline{c_j}) < 1.00$. Therefore, rule $Y \to \overline{c_j}$ is never the strongest one (since rule $X \to c_j$ is always stronger than it), an thus $I_S[f_S] = 0$.

$\square$

Theorem 3.2 gives a lower bound for σ_{min} which ensures that $I_S[f_S] = 0$. Unfortunately, in real-world scenarios, setting σ_{min} to $\frac{1}{|S|}$, is not practical, since the number of decision rules that will be generated is overwhelming. In practice, as will be empirically shown in Sect. 3.4.2, $I_S[f_S] \to 0$ when $\sigma_{min} \to \frac{1}{|S|}$. Next we present an example which illustrates the basic steps of EAC-SR.

Example Consider Table 3.1. There are ten pairs in S. All inputs were discretized. Suppose σ_{min} is set to 0.30. In this case:

- F_1 contains:

  1. $\{a_1 = [0.00-0.22] \to \text{output} = 1\}(\theta = 0.75)$
  2. $\{a_2 = [0.33-0.71] \to \text{output} = 1\}(\theta = 0.80)$
  3. $\{a_3 = [0.00-0.35] \to \text{output} = 1\}(\theta = 1.00)$
  4. $\{a_3 = [0.35-1.00] \to \text{output} = 0\}(\theta = 0.60)$

- F_2 contains:

  1. $\{a_1 = [0.00-0.22] \land a_2 = [0.33-0.71] \to \text{output} = 1\}(\theta = 0.75)$
  2. $\{a_1 = [0.00-0.22] \land a_3 = [0.00-0.35] \to \text{output} = 1\}(\theta = 1.00)$
  3. $\{a_2 = [0.33-0.71] \land a_3 = [0.00-0.35] \to \text{output} = 1\}(\theta = 1.00)$

- F_3 contains:

  1. $\{a_1 = [0.00-0.22] \land a_2 = [0.33-0.71] \land$
     $a_3 = [0.00-0.35] \to \text{output} = 1\}(\theta = 0.75)$

- $F_4 = \emptyset$, and no more frequent decision rules can be extracted from S.

Clearly, $R = \{F_1 \cup F_2 \cup F_3\}$. There is one input in T, x_{11}, for which the corresponding output, y_{11}, is unknown. The selected function f_S will use the rule set $R^{x_{11}}$ in order to predict such output. $R^{x_{11}}$ contains only 2 rules:

1. $\{a_2 = [0.33-0.71] \rightarrow output = 1\}(\theta = 0.80)$
2. $\{a_3 = [0.35-1.00] \rightarrow output = 0\}(\theta = 0.60)$

According to Eq. 3.3, EAC-SR simply picks the strongest rule in $R^{x_{11}}$, and thus it predicts output 1 for input x_{11}. In the remaining of this book we propose several improvements to EAC-SR. Such improvements lead to other associative classification algorithms, which are more efficient and practical. Thus, EAC-SR is a starting point for the associative classification algorithms discussed in this book.

Algorithm EAC-MR A single decision rule is simply a local mapping of parts of some inputs to an output, and thus it only provides a fragmented, incomplete information about $P(y|x)$. This makes EAC-SR prone to error, since it picks a single, very strong decision rule to perform predictions. Such a simple pick may provide biased predictions, and is likely to suffer from overfitting (very strong rules tend to be too specific). A safer strategy is to produce a global mapping by combining the predictions of multiple decision rules, so that a collective prediction can be performed. Intuitively, that would help avoiding bias and overfitting [12].

Given input x_i, EAC-MR (standing for EAC using multiple rules) produces a mapping function f_S, which returns the output (or the class) which receives the highest score in a weighted voting process. Specifically, R^{x_i} is interpreted as a poll, in which each decision rule $X \rightarrow c_j \in R^{x_i}$ is a vote given by X for output c_j. The weight of a vote $X \rightarrow c_j$ depends on $\theta(X \rightarrow c_j)$. The score associated with output c_j for input x_i, denoted as $s(x_i, c_j)$, is:

$$s(x_i, c_j) = \frac{\sum\limits_{r \in R^{x_i}_{c_j}} \theta(r)}{|R^{x_i}_{c_j}|} \tag{3.4}$$

The likelihood of class c_j being the output of input x_i, denoted as $\hat{p}(c_j|x_i)$, is:

$$\hat{p}(c_j|x_i) = \frac{s(x_i, c_j)}{\sum\limits_{k=1}^{p} s(x_i, c_k)} \tag{3.5}$$

where p is the number of possible outputs (i.e., the number of distinct classes in S). Finally:

$$f_S(x_i) = c_j \text{ such that } \hat{p}(c_j|x_i) \text{ is argmax } (\hat{p}(c_k|x_i)), \text{ where } 1 \leq k \leq p \tag{3.6}$$

The basic steps of EAC-MR are shown in Algorithm 2.

Algorithm 2 Finding f_S, according to EAC-MR.

Require: The training data S, and σ_{min}, and l
Ensure: f_S.

1: $R \Leftarrow$ rules $\{X \to c_j\}$ extracted from S, such that $|X| \leq l$ and $\sigma(X \to c_j) \geq \sigma_{min}$
2: return f_S such that $f_S(x_i) = c_j$, where $\hat{p}(c_j|x_i)$ is $\text{argmax}(\hat{p}(c_k|x_i)) \ \forall \ 1 \leq k \leq p$

3.3.3 Function Approximation (or Model Selection)

While the main objective of EAC-SR is to learn a function f_S for which $I_S[f_S] = 0$ (i.e., consistency with training data), EAC-MR aims at learning a function f_S for which $I_S[f_S] \approx I_T[f_S]$ (i.e., generalization). While Theorems 3.1 and 3.2 state that EAC-SR produces functions that are consistent with the training data (i.e., $I_S[f_S] = 0$), EAC-MR must rely on specific function approximation strategies in order to learn functions that are more likely to generalize.

EAC-MR may employ two function approximation techniques, which were described in Sect. 2.3 This leads to two new algorithms, EAC-MR-ERM, which employs the well known Empirical Risk Minimization technique to approximate the target function, and EAC-MR-SRM, which employs the Structural Risk Minimization technique to approximate the target function. Both algorithms will be described next.

Algorithm EAC-MR-ERM A simple strategy to empirical risk minimization is to include only highly complex, very long decision rules in R. In this case, a function f_S which makes predictions according to rules in R will fit S almost perfectly. The obvious problem is overfitting, because too complex decision rules may be as long as the training data itself, and thus f_S is unlikely to generalize. A likely to generalize function, on the other hand, would use decision rules which are considerably shorter than the training data. The problem, in this case, is underfitting, since too simple decision rules may not capture the underlying properties of S. Figure 3.2 illustrates this idea. High-order polynomials are able to fit S, while low-order polynomials are unable to properly fit S. The same trend can be observed in mapping functions that use either very complex or very simple decision rules.

Selecting a function f_S with the appropriate complexity is challenging, but mandatory for effective classification. An approach to control the complexity of f_S is to establish a relationship between how well f_S fits S (i.e., how close $I_S[f_S]$ is to 0), and how much f_S is dependent on S. If variations in S (i.e., removing a pair) do not considerably change f_S, then f_S is likely to fit pairs that are not in S as well as it fits pairs that are in S (i.e., $I_S[f_S] \approx I_T[f_S]$). If, additionally, $I_S[f_S]$ is close to 0, then f_S effectively approximates $P(y|x)$. The stability [3, 11, 14] of a function, which is denoted as β, measures at which extent small variations on S change f_S.

One way to assess the stability of a function f_S is to compare $f_S(x_i)$ with $f_{S-s_i}(x_i)$ at each pair $s_i = (x_i, y_i) \in S$ (where f_{S-s_i} is the function selected from $\{S - s_i\}$, that is, the training data after the specific pair s_i is removed). According to Eq. 2.8, the stability of f_S is given by the highest value of $|f_S(x_i) - f_{S-s_i}(x_i)|$ over all pairs in S. Stability can be used to select a function with appropriate complexity.

Selecting f_S according to Stability The hypothesis space is traversed by evaluating candidate functions in increasing order of complexity. That is, simpler functions are produced before more complex ones. To achieve this, we introduce a nested structure of subsets of rules, $h_1, h_2, \ldots, h_l$, where h_i contains frequent decision rules $X \rightarrow c_j$ for which $|X| \leq l$. Clearly, $h_1 \subseteq h_2 \subseteq \cdots \subseteq h_l$. The simplest candidate function is the one which uses decision rules in h_1 (i.e., this function uses only decision rules of size 1), while the most complex one uses decision rules in h_l. From these candidate functions, we select the one which minimizes Eq. 2.9. This selection process involves a trade-off: as the complexity of the rules increases the empirical error decreases, but the stability of the corresponding function also decreases. Figure 3.3 illustrates this trade-off. It shows functions obtained before and after removing the black point. Figure in the left shows the case in which f_S is stable (i.e., low variation between f_S and f_{S-s_i} where s_i is the black point) but it does not fit S well. Figure in the middle shows the desirable case, in which f_S fits S relatively well and is stable. Figure in the right shows the case in which f_S fits S very well, but it is not stable anymore. The selected function is the one that best trades empirical error and stability. Algorithm 3 shows basic steps of EAC-MR-ERM.

Algorithm 3 Finding f_S, according to EAC-MR-ERM.

Require: The training data S, σ_{min}, l, and δ
Ensure: f_S.

1: *tighest bound* $\Leftarrow \infty$
2: **for** $i = 1$ to l **do**
3:     $R \Leftarrow h_i \Leftarrow$ rules $\{X \rightarrow c_j\}$, such that $|X| \leq i$ and $\sigma(X \rightarrow c_j) \geq \sigma_{min}$
4:     **for** each pair $s_i = (x_i, y_i) \in S$ **do**
5:         $\beta_{s_i} = |f_S(x_i) - f_{S-s_i}(x_i)|$
6:     **end for**
7:     $\beta = \sup(\beta_{s_i})$
8:     *bound* $\Leftarrow I_S[f_S] + \left(\beta + (4n\beta + 1) \times \sqrt{\frac{\ln \frac{1}{\delta}}{2n}} \right)$
9:     **if** *tighest bound* $\leq$ *bound* **then** break
10:     *tighest bound* $\Leftarrow$ *bound*
11: **end for**
12: **return** f_S such that $f_S(x_i) = c_j$, where $\hat{p}(c_j|x_i)$ is $\mathrm{argmax}(\hat{p}(c_k|x_i)) \ \forall \ 1 \leq k \leq p$

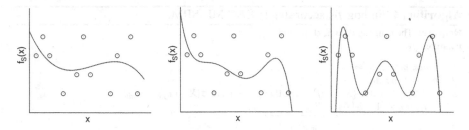

Fig. 3.2 Polynomials of increasing degrees

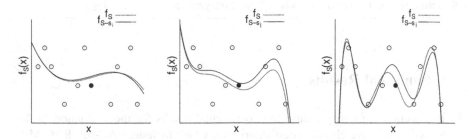

Fig. 3.3 Trading-off complexity and stability

Algorithm EAC-MR-SRM Another way of controlling the complexity of f_S is to establish a relationship between how well f_S fits S, and how complex is f_S. In Fig. 3.2, where the x-axis represents the input space, and the y-axis represents the output space, the complexity of a function is given by the number of free parameters (i.e., the polynomial degree). A more general measure of the complexity of a mapping function f_S is its VC-dimension, denoted as d_{f_S}. Ideally, f_S has small VC-dimension and $I_S[f_S]$ is close to 0.

> **Selecting f_S according to __VC-dimension__.** The hypothesis space is traversed by evaluating candidate functions in increasing order of complexity. That is, simpler functions are produced before more complex ones. Again, it is employed a nested structure of subsets of rules, $h_1, h_2, \ldots, h_l$, where h_i contains frequent decision rules $X \to c_j$ for which $|X| \leq l$. From these candidate functions, the selected is the one which minimizes Eq. 2.10. This selection process involves a trade-off: as the complexity of the rules increases the empirical error decreases, but the VC-dimension of the corresponding function increases. The selected function is the one that best trades empirical error and VC-dimension. Algorithm 4 shows the detailed steps of EAC-MR-SRM.

Algorithm 4 Finding f_S, according to EAC-MR-SRM.

Require: The training data S, σ_{min}, and δ
Ensure: f_S.

1: *tighest bound* $\Leftarrow \infty$
2: **for** $i = 1$ to l **do**
3: $\quad R \Leftarrow h_i \Leftarrow$ rules $\{X \to c_j\}$, such that $|X| \leq i$ and $\sigma(X \to c_j) \geq \sigma_{min}$
4: $\quad d_{f_S} = n \times (1 - I_S[f_S])$
5: $\quad bound \Leftarrow I_S[f_S] + \sqrt{\dfrac{d_{f_S}\left(\ln \frac{2n}{d_{f_S}} + 1\right) - \ln \frac{\delta}{4}}{n}}$
6: $\quad$ **if** *tighest bound* $\leq bound$ **then break**
7: $\quad$ *tighest bound* $\Leftarrow bound$
8: **end for**
9: return f_S such that $f_S(x_i) = c_j$, where $\hat{p}(c_j|x_i)$ is $\text{argmax}(\hat{p}(c_k|x_i)) \ \forall \ 1 \leq k \leq p$

3.4 Empirical Results

In this section we will present the experimental results for the evaluation of the proposed associative classification algorithms, which include: EAC-SR, EAC-MR, EAC-MR-ERM, and EAC-MR-SRM.

Setup Continuous attributes were discretized using the MDL-based entropy minimization method [7], which was described in Sect. 3.1.1. In all experiments we used ten fold cross-validation and the final results of each experiment represent the average of the ten runs. All results to be presented were found statistically significant based on a t-test at 95% confidence level.

Computational Environment The experiments were performed on a Linux-based PC with a Intel Pentium III 1.0 GHz processor and 1 GB RAM.

3.4.1 The UCI Benchmark

The UCI benchmark[1] provides a method for comparing the classification performance of various classification algorithms. We used a set of 26 datasets, obtained from various different applications. In this section we will evaluate the proposed algorithms using these datasets.

Baselines The evaluation is based on a comparison involving a decision tree algorithm (C4.5), a Naive Bayes algorithm (NB), and an SVM algorithm. For C4.5 and NB algorithms, we used the corresponding implementations available in MLC++ (Machine Learning Library in C++) proposed in [9]. For SVM, we used the

---

[1] Available at http://archive.ics.uci.edu/ml/

implementation available at http://svmlight.joachims.org/ (version 3.0). The well-known CBA associative classification algorithm [13] was also used as baseline. Its implementation is available at http://www.comp.nus.edu.sg/dm2/ (version 2.1).

Parameters The implementation for Naive Bayes is non-parametric. For C4.5, we used the C4.5-auto-parm tool, available in MLC++, for finding optimum parameters for each dataset. For SVM, we used the grid parameter search tool in LibSVM [6] for finding the parameters for each dataset. For CBA, EAC-SR, EAC-MR, EAC-MR-ERM, and EAC-MR-SRM, we set $\sigma_{min} = 0.01$ (we tried some values and selected the one which yields the best average performance using cross-validation on each training fold).

Evaluation Criteria Classification performance is expressed using the conventional true error rate in the test set.

Analysis Table 3.2 shows classification performance for different classification algorithms. Best results, including statistical ties, are shown in bold. EAC-SR outperforms C4.5 and NB. In fact, it was demonstrated in [17] that, if we set all algorithms under the information gain principle, an associative classification algorithm always outperforms the corresponding decision tree one. EAC-MR outperforms all baselines, showing the importance of employing multiple decision rules while approximating the target function. Furthermore, EAC-MR-ERM and EAC-MR-SRM were the best performers, suggesting that more sophisticated function approximation strategies are also capable of improving classification performance.

3.4.2 Document Categorization

Organizing documents in order to facilitate fast access to information is a long-existing necessity. In the ancient times, the Library of Alexandria contained more than 120,000 scrolls, and, at that time, finding the desired information could take days. *Callimachus of Cyrene* is considered the first bibliographer, and is the one who proposed to organize the scrolls in the library by their subjects. He invented the first library catalog, a catalog of scrolls, which revolutionized the way people store information.

Currently, fast access to information is still a central issue in digital libraries. In order to build the catalog, it is necessary to effectively group documents by common topics or subjects—a task known as document categorization. The dominant approach to document categorization is based on the application of classification algorithms. In this case, for a document given as input, its subject (or category) must be given as output.

In this section we will evaluate the proposed algorithms using a collection of documents extracted from the ACM digital library.[2] The collection contains 6,682

---

[2] http://portal.acm.org/dl.cfm/

Table 3.2 Classification performance for different algorithms

Dataset	EAC				Baselines			
	SR	MR	MR-ERM	MR-SRM	C4.5	NB	SVM	CBA
Anneal	**0.025**	**0.025**	**0.025**	**0.025**	0.075	**0.027**	0.051	**0.021**
Australian	**0.146**	**0.138**	**0.138**	**0.138**	0.148	**0.140**	**0.143**	**0.146**
Auto	0.176	**0.156**	**0.156**	**0.156**	0.176	0.321	0.249	0.199
Breast	0.106	0.074	0.074	0.074	0.056	**0.024**	**0.028**	0.037
Cleve	0.142	**0.132**	**0.132**	**0.129**	0.215	0.171	0.166	0.171
Crx	**0.127**	**0.127**	**0.123**	**0.127**	0.150	0.146	0.144	0.146
Diabetes	0.261	0.254	0.244	**0.230**	0.261	0.244	**0.230**	0.255
German	0.265	0.256	**0.247**	**0.247**	0.284	**0.246**	0.288	0.265
Glass	**0.255**	**0.255**	0.253	0.251	0.304	0.294	0.291	0.261
Heart	0.168	**0.144**	**0.144**	**0.144**	0.218	0.181	**0.142**	0.181
Hepatitis	0.183	0.183	0.167	**0.150**	0.182	**0.150**	0.167	0.189
Horse	0.176	0.176	0.176	0.176	**0.147**	0.206	0.178	0.176
Hypo	0.062	0.047	0.047	0.040	**0.007**	**0.015**	**0.013**	**0.001**
Ionosphere	0.077	**0.063**	**0.063**	**0.063**	0.105	0.119	0.083	0.077
Iris	0.081	**0.053**	0.067	**0.053**	**0.047**	0.060	**0.043**	**0.053**
Labor	**0.033**	**0.033**	**0.033**	**0.033**	0.223	0.140	0.218	0.137
Led7	0.325	0.281	0.267	**0.252**	0.305	0.267	**0.252**	0.281
Lymph	0.180	**0.130**	**0.130**	**0.130**	0.238	0.244	0.197	0.221
Pima	0.301	0.258	0.245	**0.230**	0.258	0.245	0.230	0.271
Sick	0.061	0.061	0.061	0.061	**0.011**	0.039	0.032	0.028
Sonar	0.220	**0.067**	**0.067**	**0.067**	0.284	0.230	0.160	0.225
Tic–tac–toe	0.138	0.108	0.100	0.100	0.138	0.301	0.167	**0.004**
Vehicle	0.315	0.315	0.310	0.310	0.285	0.401	**0.254**	0.310
Waveform	0.225	0.225	0.219	0.219	0.228	0.193	**0.102**	0.203
Wine	0.050	0.050	**0.00**	**0.00**	0.073	0.095	0.021	0.050
Zoo	0.096	0.096	0.070	0.070	0.078	0.137	0.049	**0.032**
Avg	0.161	0.142	**0.137**	**0.134**	0.173	0.178	0.150	0.151

documents, which were labeled using 8 first level categories of ACM, namely: Hardware, Computer Systems Organization, Software, Computing Methodologies, Mathematics of Computing, Information Systems, Theory of Computation, Computing Milieux. Citations and words in title/abstract of a document compose the corresponding set of features. The collection has a vocabulary of 9,840 unique words, and a total of 51,897 citations.

Baselines The evaluation is based on a comparison involving general-purpose algorithms, such as kNN and SVM. For kNN, we used the implementation available in MLC++ [9]. For SVM, we used the implementation available at http:// svmlight.joachims.org/ (version 3.0). Application-specific methods were also used in the evaluation. Amsler [2] is a very used bibliographic-based method for document categorization. The Bayesian method [4] and Multi-Kernel SVMs [8] are two state-of-the-art representatives of methods for document categorization.

Parameters The implementation for Amsler and Bayesian methods are non-parametric. For SVM and Multi-Kernel SVMs, we used the grid parameter search tool

Table 3.3 Categorization performance for different algorithms

Algorithms	MicF$_1$	MacF$_1$	Execution Time
Amsler (baseline)	0.832	0.783	1,251 s
EAC-MR	0.766	0.692	2,350 s
EAC-MR-ERM	0.789	0.736	2,921 s
EAC-MR-SRM	0.812	0.767	2,419 s
kNN	0.833	0.774	**83** s
SVM	0.845	**0.810**	1,932 s
Bayesian	0.847	0.796	8,281 s
Multi-Kernel	**0.859**	**0.812**	14,894 s

in LibSVM [6] for finding the optimum parameters. For kNN, we carefully tuned $k = 15$. For EAC-MR, EAC-MR-ERM, and EAC-MR-SRM, we set $\sigma_{min} = 0.005$.
Evaluation Criteria Categorization performance for the various methods being evaluated is expressed through F$_1$ measures. In this case, precision p is defined as the fraction of correctly classified documents in the set of documents classified as positive. Recall r is defined as the fraction of correctly classified documents out of all the documents having the target category. F$_1$ is a combination of precision and recall defined as the harmonic mean $\frac{2pr}{p+r}$. Macro- and micro-averaging [18] were applied to F$_1$ to get single performance values over all classification tasks. For F$_1$ macro-averaging (MacF$_1$), scores were first computed for individual categories and then averaged over all categories. For F$_1$ micro-averaging (MicF$_1$), the decisions for all categories were counted in a joint pool. The computational efficiency is evaluated through the total execution time, that is, the processing time spent in training and classifying all documents.
Analysis Table 3.3 shows categorization performance for various classification algorithms. Amsler was used as the baseline for comparison. The Multi-Kernel algorithm achieved the highest values of MicF$_1$ and MacF$_1$. Associative classification algorithms were not very effective in this application scenario, mainly due to the large number of features, which leads to a huge number of decision rules.

We analyzed the discrepancy between rule confidence values in S and T. If the discrepancy is low, then the algorithm is likely to be effective. Figure 3.4 shows the average discrepancy in confidence values according to the support. Clearly, rules with higher support values are more reliable.

At first glance, one might expect that rules with lower values of support are not reliable, and should therefore be discarded. This is not always true, as shown in Fig. 3.5, which shows the relationship between σ_{min}, execution time, and MicF$_1$. As it can be seen, higher MicF$_1$ numbers are obtained when lower σ_{min} values are employed. This is because some documents in the test set contain rare features, and consequently, such documents may demand rules with low support values. Although the information provided by such rules may not be reliable, having this information is still better than having no information at all. The problem, however, is that the execution time increases exponentially as σ_{min} decreases. This happens because, an overwhelming number of decision rules are extracted from

Fig. 3.4 Average discrepancy of rule confidence as a function of rule support

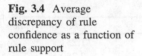

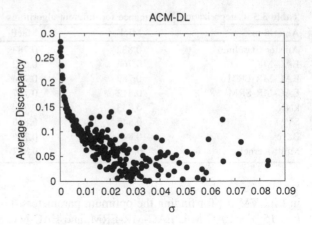

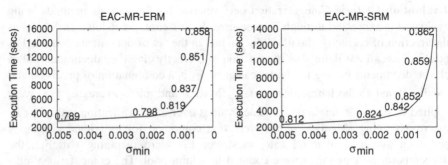

Fig. 3.5 Relationship between σ_{min}, MicF$_1$, and execution time

S. This problem may prevent associative classification algorithms to achieve full potential. What is needed is a classification algorithm which is able to extract only indispensable rules, without incurring unnecessary overhead (even for lower values of σ_{min}). In the next chapter, we will present this algorithm.

References

1. Agrawal, R., Imielinski, T., Swami, A.: Mining association rules between sets of items in large databases. In: Proceedings of the International Conference on Management of Data (SIGMOD), pp. 207–216. ACM Press (1993)
2. Amsler, R.: Application of citation-based automatic classification. Technical report, The University of Texas at Austin, Linguistics Research Center (1972)
3. Bousquet, O., Elisseeff, A.: Stability and generalization. J. Mach. Learn. Res. **2**, 499–526 (2002)
4. Calado, P., Cristo, M., Moura, E., Ziviani, N., Ribeiro-Neto, B., Gonçalves, M.: Combining link-based and content-based methods for web document classification. In: Proceedings of the Conference on Information and Knowledge Management (CIKM), pp. 394–401. ACM Press (2003)

5. Chaitin, G.: In the length of programs for computing finite binary sequences: Statistical considerations. J. ACM **16**, 145–159 (1969)
6. Chang, C.-C., Lin, C.-J.: LIBSVM: a library for support vector machines (2001). Available at http://www.csie.ntu.edu.tw/ ~ cjlin/papers/libsvm.pdf
7. Fayyad, U., Irani, K.: Multi interval discretization of continuous-valued attributes for classification learning. In: Proceedings of the International Joint Conference on Artificial Intelligence (IJCAI), pp. 1022–1027. (1993)
8. Joachims, T., Cristianini, T., Shawe-Taylor, J.: Composite kernels for hypertext categorisation. In: Proceedings of the International Conference on Machine Learning (ICML), pp. 250–257. ACM Press (2001)
9. Kohavi, R., Sommerfield, D., Dougherty, J.: Data mining using MLC++: a machine learning library in C++. In: Tools with Artificial Intelligence, pp. 234–245. (1996)
10. Kolmogorov, A.: Three approaches to the quantitative definition of information. Prob. Inform. Transm. **1**, 4–7 (1965)
11. Kutin, S., Niyogi, P.: Almost-everywhere algorithmic stability and generalization error. In: Proceedings of the Conference on Uncertainty in Artificial Intelligence (UAI), pp. 275–282. (2002)
12. Li, W., Han, J., Pei, J.: Efficient classification based on multiple class-association rules. In: Proceedings of the International Conference on Data Mining (ICDM), pp. 369–376. IEEE Computer Society (2001)
13. Liu, B., Hsu, W., Ma, Y.: Integrating classification and association rule mining. In: Proceedings of the Conference on Data Mining and Knowledge Discovery (KDD), pp. 80–86. ACM Press (1998)
14. Mukherjee, S., Niyogi, P., Poggio, T., Rifkin, R.: Learning theory: stability is sufficient for generalization and necessary and sufficient for consistency of empirical risk minimization. Adv. Comput. Math. **25**(1–3), 161–193 (2006)
15. Rissanen, J.: Modeling by shortest data description. Automatica **14**, 465–471 (1978)
16. Solomonoff, R.: A formal theory of inductive inference. Inform. Control **7**, 1–22 (1964)
17. Veloso, A., Meira, W. Jr., Zaki, M. J.: Lazy associative classification. In: Proceedings of the International Conference on Data Mining (ICDM), pp. 645–654. IEEE Computer Society (2006)
18. Yang, Y., Slattery, S., Ghani, R.: A study of approaches to hypertext categorization. J. Intell. Inf. Sys. **18**(2–3), 219–241 (2002)

Chapter 4
Demand-Driven Associative Classification

Abstract The ultimate goal of classification algorithms is to achieve the best possible classification performance for the problem at hand. Most often, classification performance is obtained by assessing some accuracy criterion using the test set, T. Therefore, an effective classification algorithm does not need to approximate the target function over the entire input space (i.e., over all possible inputs). Rather, it needs only to approximate the parts of the target function that are defined over the inputs in T. As discussed, it is often hard to approximate the target function defined over inputs in T, using a single mapping function. The key insight is to produce a specifically designed function, $f_S^{x_i}$, which approximates the target function at each input $x_i \in T$. Thus, a natural way to improve classification performance is to predict the outputs for inputs in T, on a demand-driven basis. In this case, particular characteristics of each input in T may be taken into account while predicting the corresponding output. The expected result is a set of multiple mapping functions, where each function $f_S^{x_i}$ is likely to perform particularly accurate predictions for input $x_i \in T$, no matter the (possible poor) performance in predicting outputs for other inputs. In this chapter we introduce methods and algorithms for demand-driven associative classification.

Keywords Lazy association classification · Projection · Demand-driven rule extraction · Pruning · Caching · Demand-driven function approximation · Document categorization

4.1 Method and Algorithms for Demand-Driven Associative Classification

In this section we will present a method for associative classification, which produces mapping functions on a demand-driven basis, according to each input in T.

A. Veloso and W. Meira Jr., *Demand-Driven Associative Classification*,
SpringerBriefs in Computer Science, DOI: 10.1007/978-0-85729-525-5_4,
© Adriano Veloso 2011

First, it will be described how decision rules are extracted from S, and then how these rules are used in order to approximate $P(y|x)$. The method to be presented will hereafter be referred to as LAC (standing for Lazy Associative Classification). Algorithms based on this method will also be presented in this section.

4.1.1 Correlation Preserving Projection

In demand-driven associative classification, whenever an input $x_i \in T$ is being considered, that input is used as a filter to remove from S, features and examples that are useless[1] to approximate the target function at input x_i. This process generates a *projected training data*, defined as S^{x_i}. Specifically, given an input $x_i \in T$, the corresponding projected training data, S^{x_i}, is the set of examples in S, which is obtained after removing all features that are not included in x_i.

Theorem 4.1. *Given an input $x_i \in T$, a decision rule $X \rightarrow c_j$, such that $X \subseteq x_i$, has the same confidence value in S and S^{x_i}.*

Proof. Comes directly from the fact that confidence is null-invariant, thus adding or removing examples that do not contain X, does not change $\theta(X \rightarrow c_j)$.   □

Theorem 4.1 states that the projection preserves the original correlation between inputs and outputs, since confidence values for rules extracted from S and S^{x_i} have the same value.

4.1.2 Demand-Driven Rule Extraction

Rule extraction is a major issue when devising an associative classification algorithm. Typically, frequent rules are extracted from S, and used to produce a mapping function. Two problems may arise: (i) an infrequent, but useful rule may be not extracted from S (possibly hurting classification performance), and (ii) a frequent, but useless rule may be extracted from S (incurring unnecessary overhead). Figure 4.1 illustrates these problems. In the figure, black points represent useful rules, and white points represent useless rules. σ_{min} induces a border which separates frequent from infrequent rules. If σ_{min} is set too high, few useless rules are extracted from S, but several useful rules are not extracted. If σ_{min} is set too low, all useful rules are extracted from S, but a prohibitive number of useless rules is also extracted. An optimal value for σ_{min}, in the sense that only useful rules are extracted from S, is unlikely to exist.

---

[1] Usefulness is defined in Sect. 3.2.

Fig. 4.1 The pruning dilemma

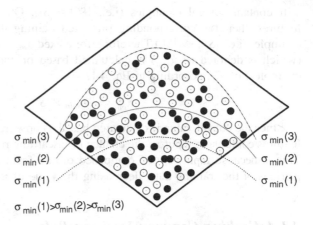

$\sigma_{min}(3)$ $\sigma_{min}(3)$
$\sigma_{min}(2)$ $\sigma_{min}(2)$
$\sigma_{min}(1)$ $\sigma_{min}(1)$

$\sigma_{min}(1) > \sigma_{min}(2) > \sigma_{min}(3)$

An ideal scenario would be to extract only useful rules from S, without discarding useful ones. Inputs in T have valuable information that can be used during rule extraction to guide the search for useful rules. In this section, we propose to achieve the ideal scenario by extracting rules from S on a demand-driven basis, according to the input in T being considered.

Theorem 4.2. *Given an input $x_i \in T$, R^{x_i} contains only decision rules that are useful to approximate the target function at x_i.*

Proof. Rules in R^{x_i} are extracted from S^{x_i}. Since all inputs in S^{x_i} contain only features that are included in x_i, the existence of a rule $X \rightarrow c_j \in R^{x_i}$, such that $X \nsubseteq x_i$, is impossible.   □

Thus, according to Theorem 4.2, only useful rules are extracted from S. Next, we will discuss more sophisticated pruning strategies that reduce the chance of discarding useful rules.

4.1.3 Demand-Driven Pruning with Multiple Cut-Off Values

As discussed in the previous chapter, the typical pruning strategy is based on a cut-off frequency value derived from σ_{min}, which separates frequent from infrequent rules. In many cases, however, rules with low support may carry useful information for the sake of prediction. In such cases, the "use-and-abuse" support-based pruning strategy [2] is not appropriate, since useful rules may also be discarded. We propose an alternate strategy, which prevents support-based pruning from being excessive. The proposed strategy employs multiple cut-off values. Specifically, cut-off values are calculated depending on how frequent (or how rare) are the features composing input $x_i \in T$. The key insight is that if x_i contains commonly-appearing features, then the corresponding projected training data, S^{x_i}, will contain several examples (i.e., $|S^{x_i}| \approx |S|$). Otherwise, if x_i contains rare features, then the corresponding projected training data will contain only few examples (i.e., $|S^{x_i}| \ll |S|$.) Therefore, for a fixed σ_{min}. the cut-off value for input x_i (which is denoted as $\pi_{min}^{x_i}$) is calculated based on the size (i.e., the number of examples) of S^{x_i}, according to Eq. 4.1.

$$\pi_{min}^{x_i} = \lceil \sigma_{min} \times |S^{x_i}| \rceil \qquad (4.1)$$

Since the cut-off value is calculated based on how frequent (or how rare) are the features composing input x_i, the chance of discarding rules matching x_i is reduced. This is because inputs $x_i \in T$ composed of rare features will produce small projections of the training data, decreasing the value of $\pi_{min}^{x_i}$.

4.1.4 Caching Common Decision Rules

Extracting a decision rule has a significant computational cost, since it involves accessing S multiple times. Different inputs in T may demand the extraction of different rules, but it is very likely that some of these rules are common. In this case, caching (or memorization) is very effective in reducing work replication and, consequently, to reduce the number of accesses to S.

Our cache is a pool of entries, and each entry has the form $\langle key, data \rangle$, where $key = \{X, c_j\}$ and $data = \{\sigma(X \rightarrow c_j), \theta(X \rightarrow c_j)\}$. Our cache implementation has a limited storage and stores all cached decision rules in main memory. Before extracting rule $X \rightarrow c_j$, algorithms based on the LAC method first check whether this rule is already in the cache. If an entry is found with a key matching $\{X, c_j\}$, then the rule in the cache entry is used instead of extracting it from S. If it is not found, the rule is extracted from S and then it can be inserted into the cache.

The cache size is limited, and when the cache is full, some rules must be discarded to make room for other ones. The replacement heuristic is based on the support of the rules. More specifically, the least frequent rule stored in the cache is the first one to be discarded (and it will only be discarded if the rule to be inserted is more frequent than it). There are two main reasons to adopt this heuristic. First, the more frequent is the rule, the higher is the chance of using this rule later, to predict the output for other inputs in T (thus, if a frequent rule is discarded, it is higher the chance of recalculating it later on). Second, the computational cost associated with the extraction of more frequent rules is higher than the cost associated with the extraction of less frequent ones (more frequent rules necessitates more accesses to S).

4.1.4.1 Prediction

Mapping functions perform predictions using the decision rules extracted from S^{x_i}, as will be discussed next.

Algorithm LAC-SR This algorithm employs a single decision rule, which is the strongest one in R^{x_i} (i.e., the one with highest confidence value), in order to predict the output for input $x_i \in T$, according to Eq. 4.2.

$$f_S^{x_i}(x_i) = c_j \text{ such that } \theta(X \to c_j) \text{ is argmax } (\theta(r)) \forall r \in R^{x_i} \qquad (4.2)$$

Algorithm LAC-MR This algorithm employs multiple decision rules to produce a specific function, $f_S^{x_i}$, for input $x_i \in T$. It first projects the original training data, S, according to input x_i. The result is S^{x_i}. Then, LAC-MR extracts rules from S^{x_i}, on a demand-driven basis. All extracted rules are useful to x_i. Basic steps for LAC-MR are shown in Algorithm 5.

Algorithm 5 Finding $f_S^{x_i}$, according to LAC-MR.

Require: The training data S, input $x_i \in T$, σ_{min}, and l
Ensure: $f_S^{x_i}$.

1: $S^{x_i} \Leftarrow S$ projected according to x_i
2: $R^{x_i} \Leftarrow$ rules $X \to c_j$ extracted from S^{x_i}, such that $\pi(X \to c_j) \geq \pi_{min}^{x_i}$
3: return $f_S^{x_i}$ such that $f_S^{x_i}(x_i) = c_j$, where $\hat{p}(c_j|x_i)$ is argmax$(\hat{p}(c_k|x_i)) \forall 1 \leq k \leq p$

Example Consider the example shown in Table 4.1. There are ten pairs in S. There is one input in T, x_{12}, for which the corresponding output, y_{12}, is unknown. All inputs were discretized. After projecting S according to input x_{12}, we obtain $S^{x_{12}}$, which is shown in Table 4.2.

Suppose σ_{min} is set to 0.30. In this case, $R^{x_{12}}$ contains the following 6 rules:

1. $\{a_1 = [0.46-1.00] \to output = 1\}(\theta = 1.00)$
2. $\{a_1 = [0.46-1.00] \land a_3 = [0.35-1.00] \to output = 1\}(\theta = 1.00)$
3. $\{a_2 = [0.00-0.33] \to output = 0\}(\theta = 0.67)$
4. $\{a_2 = [0.00-0.33] \land a_3 = [0.35-1.00] \to output = 0\}(\theta = 0.67)$

Table 4.1 Training data and test set given as example

		Input (x_i)			Output (y_i)
		a_1	a_2	a_3	
	(x_1, y_1)	[0.00–0.22]	[0.33–0.71]	[0.00–0.35]	1
	(x_2, y_2)	[0.00–0.22]	[0.33–0.71]	[0.35–1.00]	0
	(x_3, y_3)	[0.46–1.00]	[0.71–1.00]	[0.35–1.00]	1
	(x_4, y_4)	[0.22–0.46]	[0.00–0.33]	[0.35–1.00]	0
S	(x_5, y_5)	[0.00–0.22]	[0.33–0.71]	[0.00–0.35]	1
	(x_6, y_6)	[0.22–0.46]	[0.33–0.71]	[0.00–0.35]	1
	(x_7, y_7)	[0.00–0.22]	[0.33–0.71]	[0.00–0.35]	1
	(x_8, y_8)	[0.22–0.46]	[0.71–1.00]	[0.00–0.35]	1
	(x_9, y_9)	[0.46–1.00]	[0.00–0.33]	[0.35–1.00]	1
	(x_{10}, y_{10})	[0.22–0.46]	[0.00–0.33]	[0.35–1.00]	0
T	(x_{12}, y_{12})	[0.46–1.00]	[0.00–0.33]	[0.35–1.00]	?(1)

Table 4.2 Projected training data: $S^{x_{12}}$

		Input (x_i)			Output (y_i)
		a_1	a_2	a_3	
	(x_2, y_2)	$\emptyset$	$\emptyset$	[0.35–1.00]	0
	(x_3, y_3)	[0.46–1.00]	$\emptyset$	[0.35–1.00]	1
$S^{x_{12}}$	(x_4, y_4)	$\emptyset$	[0.00–0.33]	[0.35–1.00]	0
	(x_9, y_9)	[0.46–1.00]	[0.00–0.33]	[0.35–1.00]	1
	(x_{10}, y_{10})	$\emptyset$	[0.00–0.33]	[0.35–1.00]	0

5. $\{a_3 = [0.35-1.00] \rightarrow output = 0\} (\theta = 0.60)$
6. $\{a_3 = [0.35-1.00] \rightarrow output = 1\} (\theta = 0.40)$

According to Eq. 3.6, LAC-MR calculates the likelihood associated with each output, which are $\hat{p}(output = 0|x_{12}) = 0.48$ and $\hat{p}(output = 1|x_{12}) = 0.52$. Thus.LAC-MR predicts output 1 for input x_{12}, which is the correct one.

4.1.4.2 Demand-Driven Function Approximation

A complex target function may be composed of simple parts. Thus, instead of approximating a complex target function using a complex mapping function (i.e., f_S) we can employ multiple simple functions (i.e., $f_S^{x_i}$). Intuitively, such simple functions are more likely to generalize than a single complex function. The appropriate complexity for each function $f_S^{x_i}$ is selected using the function approximation techniques described in Sect. 2.3. LAC-MR-ERM employs the well known Empirical Risk Minimization technique to approximate the target function, while LAC-MR-SRM employs the Structural Risk Minimization technique to approximate the target function.

LAC-MR-ERM and LAC-MR-SRM are much finer-grained than their eager counterparts EAC-MR-ERM and EAC-MR-SRM, since they are able to select a different complexity for each function $f_S^{x_i}$. In the end of the process these algorithms produce multiple mapping functions, each one with a possibly different complexity. Both algorithms will be described next.

Algorithm LAC-MR-ERM The hypothesis space induced by each projected training data, S^{x_i}, is traversed by evaluating candidate functions in increasing order of complexity. That is, simpler functions are produced before more complex ones. To do this, we introduce a nested structure of subsets of rules, $h_1, h_2, \ldots, h_l$, where h_i contains frequent decision rules $X \rightarrow c_j$ for which $|X| \leq l$. Clearly, $h_1 \subseteq h_2 \subseteq \cdots \subseteq h_l$. The simplest candidate function is the one which uses decision rules in h_1 (i.e., this function uses only decision rules of size 1), while the most complex one uses decision rules in h_l. From these candidate functions, we select the one which minimizes Eq. 2.9. This selection process involves a trade-off: as the complexity of the rules increases, the empirical error decreases, but the stability of the corresponding function also decreases. The selected function is the one that best trades empirical error and stability. Algorithm 6 shows the basic steps of LAC-MR-ERM.

Algorithm 6 Finding $f_S^{x_i}$, according to LAC-MR-ERM.

Require: The training data S, input $x_i \in T$, σ_{min}, l, and δ
Ensure: $f_S^{x_i}$.

1: *tighest bound* $\Leftarrow \infty$
2: $S^{x_i} \Leftarrow S$ projected according to x_i
3: **for** $i = 1$ to l **do**
4:    $R^{x_i} \Leftarrow h_i \Leftarrow$ rules $X \to c_j$ extracted from S^{x_i}, such that $|X| \leq i$ and $\pi(X \to c_j) \geq \pi_{min}^{x_i}$
5:    **for each** pair $s_i = (x_i, y_i) \in S^{x_i}$ **do**
6:       $\beta_{s_i} = |f_S^{x_i}(x_i) - f_{S-s_i}^{x_i}(x_i)|$
7:    **end for**
8:    $\beta = \sup(\beta_{s_i})$
9:    *bound* $\Leftarrow I_{S^{x_i}}[f_S^{x_i}] + \left(\beta + (4|S^{x_i}|\beta + 1) \times \sqrt{\frac{\ln\frac{1}{\delta}}{2|S^{x_i}|}} \right)$
10:    **if** *tighest bound* $\leq$ *bound* **then break**
11:    *tighest bound* $\Leftarrow$ *bound*
12: **end for**
13: **return** $f_S^{x_i}$ such that $f_S^{x_i}(x_i) = c_j$, where $\hat{p}(c_j|x_i)$ is $\mathrm{argmax}(\hat{p}(c_k|x_i))\ \forall\ 1 \leq k \leq p$

Algorithm LAC-MR-SRM The hypothesis space induced by each projected training data, S^{x_i}, is traversed by evaluating candidate functions in increasing order of complexity. That is, simpler functions are produced before more complex ones. Again, it is employed a nested structure of subsets of rules, $h_1, h_2, \ldots, h_l$, where h_i contains frequent decision rules $X \to c_j$ for which $|X| \leq l$. From these candidate functions, the selected is the one which minimizes Eq. 2.10. This selection process involves a trade-off: as the complexity of the rules increases the empirical error decreases, but the VC-dimension of the corresponding function increases. The selected function is the one that best trades empirical error and VC-dimension. Algorithm 7 shows the detailed steps of LAC-MR-SRM.

Algorithm 7 Finding $f_S^{x_i}$, according to LAC-MR-SRM.

Require: The training data S, input $x_i \in T$, σ_{min}, and δ
Ensure: $f_S^{x_i}$.

1: *tighest bound* $\Leftarrow \infty$
2: $S^{x_i} \Leftarrow S$ projected according to x_i
3: **for** $i = 1$ to l **do**
4:    $R^{x_i} \Leftarrow h_i \Leftarrow$ rules $X \to c_j$ extracted from S^{x_i}, such that $|X| \leq i$ and $\pi(X \to c_j) \geq \pi_{min}^{x_i}$
5:    $d_{f_S^{x_i}} = |S^{x_i}| \times (1 - I_{S^{x_i}}[f_S^{x_i}])$
6:    *bound* $\Leftarrow I_{S^{x_i}}[f_S^{x_i}] + \sqrt{\dfrac{d_{f_S^{x_i}}\left(\ln\frac{2|S^{x_i}|}{d_{f_S^{x_i}}}+1\right) - \ln\frac{\delta}{4}}{|S^{x_i}|}}$
7:    **if** *tighest bound* $\leq$ *bound* **then break**
8:    *tighest bound* $\Leftarrow$ *bound*
9: **end for**
10: **return** $f_S^{x_i}$ such that $f_S^{x_i}(x_i) = c_j$, where $\hat{p}(c_j|x_i)$ is $\mathrm{argmax}(\hat{p}(c_k|x_i))\ \forall\ 1 \leq k \leq p$

4.2 Empirical Results

In this section we will present the experimental results for the evaluation of the proposed demand-driven associative classification algorithms, which include: LAC-SR, LAC-MR, LAC-MR-ERM, LAC-MR-SRM .

Setup Continuous attributes were discretized using the MDL-based entropy minimization method [5], which was described in Sect. 3.1.1. In all the experiments we used ten-fold cross-validation and the final results of each experiment represent the average of the ten runs. All the results to be presented were found statistically significant based on a t-test at 95% confidence level.

4.2.1 The UCI Benchmark

Baselines The evaluation is based on a comparison involving a lazy decision tree algorithm (LazyDT), a kNN algorithm (kNN), and an associative classification algorithm (DeEPs). For LazyDT and kNN, we used the corresponding implementations available in MLC++ [7]. For DeEPs, we used the results available at [8].

Parameters For LazyDT, we used the C4.5-auto-parm tool, available in MLC++, for finding optimum parameters for each dataset. For kNN, the value of k was carefully hand-tuned for each dataset. For LAC-SR, LAC-MR, LAC-MR-ERM, and LAC-MR-SRM, we set $\sigma_{min} = 0.01$.

Evaluation Criteria Classification performance is expressed using the conventional true error rate in the test set.

Analysis Table 4.3 shows classification performance for different classification algorithms. Best results, including statistical ties, are shown in bold. LAC-SR outperforms LazyDT and kNN. In fact, it was demonstrated in [9] that, if we set all algorithms under the information gain principle, a demand-driven associative classification algorithm always outperforms the corresponding decision tree one. LAC-SR has also outperformed EAC-SR, providing gains of more than 8%. LAC-MR outperforms all baselines. It also outperformed EAC-MR, with gains of more than 7%. Furthermore, LAC-MR-ERM and LAC-MR-SRM were the best performers, suggesting that more finer-grained function approximation strategies are effective in improving classification performance. LAC-MR-ERM outperformed EAC-MR-ERM with gains of more than 9.5%, while LAC-MR-SRM outperformed EAC-MR-SRM with gains of more than 9.7%.

4.2.2 Document Categorization

Baselines The evaluation is based on a comparison involving general-purpose algorithms, such as kNN, SVM, and TSVM (SVM). For kNN, we used the

Table 4.3 Classification performance for different algorithms

Dataset	LAC				Baselines		
	SR	MR	MR-ERM	MR-SRM	LazyDT	kNN	DeEPs
Anneal	**0.025**	**0.025**	**0.021**	**0.021**	0.042	0.100	0.050
Australian	0.146	0.138	0.132	0.132	0.152	0.152	**0.116**
Auto	0.176	**0.156**	**0.151**	**0.151**	0.247	0.280	0.273
Breast	0.074	**0.024**	**0.021**	**0.021**	0.051	0.185	0.036
Cleve	0.142	**0.132**	**0.132**	**0.129**	0.172	0.162	0.158
Crx	**0.127**	**0.127**	**0.123**	**0.123**	0.169	0.169	**0.119**
Diabetes	**0.221**	**0.221**	**0.221**	**0.221**	0.249	0.241	0.230
German	0.265	**0.256**	**0.247**	**0.247**	0.261	**0.256**	**0.256**
Glass	0.253	0.253	**0.231**	0.243	0.265	0.372	0.326
Heart	0.168	0.144	0.144	0.137	0.177	**0.102**	0.177
Hepatitis	0.183	0.183	**0.110**	**0.110**	0.203	0.223	0.175
Horse	0.176	0.176	0.173	0.173	0.173	0.173	**0.147**
Hypo	0.062	0.047	0.032	0.032	**0.012**	**0.012**	**0.018**
Ionosphere	0.077	**0.063**	**0.063**	**0.063**	0.080	0.153	0.088
Iris	0.053	**0.033**	**0.033**	**0.033**	0.053	0.044	**0.033**
Labor	0.033	0.033	0.033	**0.016**	0.204	0.033	0.023
Led7	0.281	**0.259**	0.271	**0.263**	**0.265**	**0.263**	**0.263**
Lymph	0.180	**0.130**	**0.130**	**0.123**	0.201	0.180	0.246
Pima	0.266	**0.218**	**0.208**	**0.212**	0.259	0.279	0.229
Sick	0.061	0.061	0.061	0.026	0.021	0.094	**0.033**
Sonar	0.220	**0.067**	**0.067**	**0.067**	0.246	0.220	0.133
Tic–tac–toe	0.054	0.036	0.036	0.054	**0.006**	0.108	**0.004**
Vehicle	0.310	0.310	0.292	0.292	0.318	0.334	**0.254**
Waveform	0.225	0.225	0.212	0.196	0.225	0.225	**0.157**
Wine	0.050	**0.000**	**0.000**	**0.000**	0.079	0.263	0.039
Zoo	0.096	0.096	0.070	0.070	0.078	0.070	**0.028**
Avg	0.149	0.132	**0.124**	**0.121**	0.162	0.180	0.139

implementation available in MLC++ [7]. For SVM, we used the implementation available at http://svmlight.joachims.org/ (version 3.0). For TSVM, we used the implementation available at http://www.kyb.mpg.de/bs/people/fabee/universvm. html. Application-specific methods were also used in the evaluation. Amsler [1] is a very used bibliographic-based method for document categorization. The Bayesian method [3] and Multi-Kernel SVMs [6] are two state-of-the-art representatives of methods for document categorization.

Parameters The implementation for Amsler and Bayesian methods are nonparametric. For SVM, TSVM, and Multi-Kernel SVMs, we used the grid parameter search tool in LibSVM [4] for finding the optimum parameters. For kNN, we carefully tuned $k = 15$. For LAC-MR, LAC-MR-ERM, and LAC-MR-SRM, we set $\sigma_{min} = 0.005$.

Evaluation Criteria Categorization performance for the various methods being evaluated, is expressed through F_1 measures. In this case, precision p is defined as the proportion of correctly classified documents in the set of all documents. Recall

Table 4.4 Categorization performance for different algorithms

Algorithms	MicF$_1$	MacF$_1$	Execution time (s)
Amsler (baseline)	0.832	0.783	1,251
kNN	0.833	0.774	83
SVM	0.845	0.810	1,932
Bayesian	0.847	0.796	8,281
TSVM	0.855	0.808	17,183
Multi-Kernel	0.859	0.812	14,894
LAC-MR	0.862	0.814	257
LAC-MR-ERM	0.868	0.833	504
LAC-MR-SRM	0.873	0.839	342

Fig. 4.2 Processing time with varying cache sizes

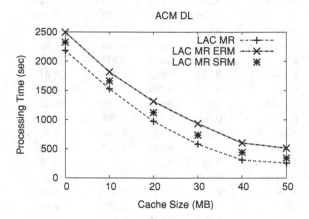

r is defined as the proportion of correctly classified documents out of all the documents having the target category. F$_1$ is a combination of precision and recall defined as the harmonic mean $\frac{2pr}{p+r}$. Macro- and micro-averaging [10] were applied to F$_1$ to get single performance values over all classification tasks. For F$_1$ macro-averaging (MacF$_1$), scores were first computed for individual categories and then averaged over all categories. For F$_1$ micro-averaging (MicF$_1$), the decisions for all categories were counted in a joint pool. The computational efficiency is evaluated through the total execution time, that is, the processing time spent in training and classifying all documents.

Analysis Table 4.4 shows categorization performance for various classification algorithms. Again, Amsler was used as the baseline for comparison. The TSVM and Multi-Kernel algorithms achieved the highest values of MicF$_1$ and MicF$_1$. LAC-MR outperformed these algorithms, showing the importance of producing mapping functions on a demand-driven basis. Finer-grained function approximation techniques were also effective. LAC-MR-ERM and LAC-MR-SRM were the best overall performers. Further, demand-driven associative classification algorithms are much faster than most of the baselines.

Figure 4.2 depicts the execution times obtained by employing different cache sizes. We varied the cache size from 0 to 50 MB, and for each storage capacity we obtained the corresponding execution time. Clearly, execution time is very sensitive to cache size. Caches as large as 50 MB are able to store all decision rules with no need of replacement, being the best cache configuration.

References

1. Amsler, R.: Application of citation-based automatic classification. Technical report, The University of Texas at Austin, Linguistics Research Center (1972)
2. Baralis, E., Chiusano, S., Garza, P.: On support thresholds in associative classification. In: Proceedings of the Symposium on Applied Computing (SAC), pp. 553–558. ACM Press (2004)
3. Calado, P., Cristo, M., Moura, E., Ziviani, N., Ribeiro-Neto, B., Gonçalves, M.: Combining link-based and content-based methods for web document classification. In: Proceedings of the Conference on Information and Knowledge Management (CIKM), pp. 394–401. ACM Press (2003)
4. Chang, C.-C., Lin, C.-J.: LIBSVM: a library for support vector machines, 2001. Available at http://www.csie.ntu.edu.tw/ ∼ cjlin/papers/libsvm.pdf
5. Fayyad, U., Irani, K.: Multi interval discretization of continuous-valued attributes for classification learning. In: Proceeding of the International Joint Conference on Artificial Intelligence (IJCAI). pp. 1022–1027 (1993)
6. Joachims, T., Cristianini, N., Shawe-Taylor, J.: Composite kernels for hypertext categorisation. In: Proceeding of the International Conference on Machine Learning (ICML), pp. 250–257. ACM Press (2001)
7. Kohavi, R., Sommerfield, D., Dougherty, J.: Data mining using MLC++: A machine learning library in C++. In: Tools with Artificial Intelligence, pp. 234–245 (1996)
8. Li, J., Dong, G., Ramamohanarao, K., Wong, L.: Deeps: A new instance-based lazy discovery and classification system. Mach. Learn. **54**(2), pp. 99–124 (2004)
9. Veloso, A.., Meira, W., Jr., Zaki, M.J.: Lazy associative classification. In: Proceeding of the International Conference on Data Mining (ICDM), pp. 645–654. IEEE Computer Society (2006)
10. Yang, Y., Slattery, S., Ghani, R. A study of approaches to hypertext categorization. J. Intell. Inf. Syst. **18**(2–3), 219–241(2002)

Part III
Extensions to Associative Classification

Chapter 5
Multi-Label Associative Classification

Abstract A typical assumption in classification is that outputs are mutually exclusive, so that an input can be mapped to only one output (i.e., single-label classification). However, due to ambiguity or multiplicity, it is quite natural that many applications violate this assumption, allowing inputs to be mapped to multiple outputs simultaneously. Multi-label classification is a generalization of single-label classification, and its generality makes it much more difficult to solve. Despite its importance, research on multi-label classification is still lacking. Common approaches simply learn independent functions (Brinker et al. Unified model for multilabel classification and ranking. In: Proceedings of the European Conference on Artificial Intelligence, ECAI, pp. 489–493, 2006), not exploiting dependencies among outputs (Boutell et al. Learning multi-label scene classification. Pattern Recogn. 37(9), 1757–1771, 2004; Clare and King, Knowledge discovery in multi-label phenotype data. In: Proceedings of the European Conference on Principles and Practice of Knowledge Discovery in Databases (PKDD), Springer, pp. 42–53, 2001). Also, several small disjuncts may appear due to the possibly large number of combinations of outputs, and neglecting these small disjuncts may degrade classification performance (Proceedings of the International Conference on Information and Knowledge Management, CIKM, 2005; Proceedings of the Conference on Computer Vision and Pattern Recognition, 2006). In this chapter we extend demand-driven associative classification to multi-label classification.

Keywords Multi-label classification · Ambiguity · Multiplicity · Independent outputs · Correlated outputs · Label focusing · Document categorization · Gene functional analysis

A. Veloso and W. Meira Jr., *Demand-Driven Associative Classification*, 53
SpringerBriefs in Computer Science, DOI: 10.1007/978-0-85729-525-5_5,
© Adriano Veloso 2011

5.1 Algorithms for Multi-Label Associative Classification

Next we present two algorithms for multi-label associative classification. The first one, which will be referred to as LAC-MR-IO (standing for LAC-MR with independent outputs), neglects any association among outputs while producing classification functions. The second one, which will be hereafter referred to as LAC-MR-CO (standing for LAC-MR with correlated outputs), explores correlated outputs, improving classification performance.

Algorithm LAC-MR-IO The assumption of independence of outputs leads to a natural strategy for multi-label classification, where class membership probabilities, $\hat{p}(c_i|x)$, are naturally used to select outputs. A user specified threshold, Δ_{min} ($0 \leq \Delta_{min} \leq 0.5$), is used to separate the outputs that will be predicted. Specifically, for a given input x, an output c_i is only predicted if $\hat{p}(c_i|x) \geq \Delta_{min}$. LAC-MR-IO follows this strategy, and the basic steps are shown in Algorithm 8.

Algorithm 8 Finding $f_S^{x_i}$, according to LAC-MR-IO.

Require: The training data S, input $x_i \in T$, σ_{min}, and Δ_{min}
Ensure: $f_S^{x_i}$.

1: $S^{x_i} \Leftarrow S$ projected according to x_i
2: $R^{x_i} \Leftarrow$ rules $X \rightarrow c_j$ extracted from S^{x_i}, such that $\pi(X \rightarrow c_j) \geq \sigma_{min} \times |S^{x_i}|$
3: return $f_S^{x_i} = P^{x_i}$, such that $\forall c_j \in P^{x_i}$, $\hat{p}(c_j|x_i) \geq \Delta_{min}$

Algorithm LAC-MR-CO Outputs in multi-label problems are often correlated, and as it will be shown in the experiments, this correlation may be helpful for improving classification performance. Next we will present LAC-MR-CO, which, unlike LAC-MR-IO, explicitly exploits interactions among outputs while producing classification functions. These functions are composed of *multi-label decision rules*, which are rules of the form $\{X \cup F\} \rightarrow c_j$, where $F \subseteq \{C - c_j\}$. Thus, multi-label decision rules enable the presence of outputs in the antecedent.

Classification functions are produced iteratively, following a greedy heuristic called *progressive label focusing* [3, 5], which tries to find the best combination of outputs by making locally best choices. In the first iteration, $F = \emptyset$, and a set of rules matching input $x \in T$, R^{x^1} (which is composed of rules of the form $X \rightarrow c_j$), is extracted from S^x. Based on R^{x^1}, output c_r is predicted for input x. In the second iteration, output c_r is treated as a new feature while extracting rules (i.e., $F = \{c_r\}$). A set of multi-label rules, R^{x^2} (which is composed of rules of the form $\{X \cup \{c_r\}\} \rightarrow c_j$, with $j \neq r$), is extracted from S^x. Based on R^{x^2}, output c_s is predicted for input x. This process iterates until no more rules are extracted from S^x. The basic idea is to progressively narrow the search space for rules as outputs are being predicted for input x. The main steps of LAC-MR-CO are shown in Algorithm 9.

Algorithm 9 Finding $f_S^{x_i}$, according to LAC-MR-CO.

Require: The training data S, input $x_i \in T$, and σ_{min}
Ensure: $f_S^{x_i}$.

1: $\Omega \Leftarrow 1$
2: $F \Leftarrow \emptyset$
3: $S^{x_i} \Leftarrow S$ projected according to x_i
4: **while** true **do**
5:     $R^{x^i} \Leftarrow$ rules $\{X \cup F\} \rightarrow c_j$ extracted from S^{x_i}, such that $\pi(\{X \cup F\} \rightarrow c_j) \geq \sigma_{min} \times |S^{x_i}|$
6:     **if** $R^{x^i} = \emptyset$ **then** break
7:     $F \Leftarrow F \cup c^\Omega$, where $\hat{p}(c^\Omega | x_i)$ is argmax$(\hat{p}(c_k | x_i))$ with $1 \leq k \leq p$
8:     Ω++
9: **end while**
10: **return** $f_S^{x_i} = \{c^1, \dots, c^\Omega\}$

Table 5.1 Training data given as example of a multi-label problem

		Output	Input	
			Title	Actors
S	x_1	Comedy/ Romance	Forrest Gump	T. Hanks
	x_2	Drama/Romance	The Terminal	T. Hanks
	x_3	Drama/Crime	Catch Me If You Can	T. Hanks and L. DiCaprio
	x_4	Drama/Crime	The Da Vinci Code	T. Hanks
	x_5	Drama/Crime	Blood Diamond	L. DiCaprio
	x_6	Crime/Action	The Departed	L. DiCaprio and M. Damon
	x_7	Crime/Action	The Bourne Identity	M. Damon
	x_8	Action/Romance	Syriana	M. Damon
	x_9	Romance	Troy	B. Pitt
	x_{10}	Drama/Crime	Confidence	E. Burns
T	x_{11}	[Action/Crime]	Ocean's Twelve	B. Pitt and M. Damon
	x_{12}	[Crime/Drama]	The Green Mile	T. Hanks

Example Table 5.1 shows an example of the multi-label classification problem. In this case, each input corresponds to a movie, and each movie is assigned to one or more labels (i.e., outputs). This movie subject was chosen because its intuitive aspects may help to understand the ideas just discussed.

Suppose we want to predict the outputs for input x_{11}. In this case, the execution of LAC-MR-IO, with $\sigma_{min} = 0.50$ and $\Delta_{min} = 0.35$, proceeds as follows. First a set of frequent rules is extracted from $S^{x_{11}}$:

1. Actor=M. Damon→output=Action ($\theta = 1.00$)
2. Actor=M. Damon→output=Crime ($\theta = 0.67$)

Following Eq. 3.5, $\hat{p}(\text{Action}|x_{11})$=0.60, and $\hat{p}(\text{Crime}|x_{11})$=0.40. In this case, both outputs, "Action" and "Crime", are correctly predicted for input x_{11}, since $\Delta_{min} = 0.35$.

Now, suppose we want to predict the outputs for input x_{12}. In this case,the execution of LAC-MR-CO, with $\sigma_{min} = 0.50$, proceeds as follows. At the first iteration, one rule is extracted from $S^{x_{12}}$:

- Actor=T. Hanks→output=Drama($\theta = 0.75$)

Obviously, output "Drama" is predicted as the output for input x_{12}. Therefore, $F = \{Drama\}$. In the next iteration, another rule is extracted from $S^{x_{12}}$:

- Actor=T.Hanks∧output=Drama→output=Crime ($\theta = 0.50$)

Since no more rules can be extracted from $S^{x_{12}}$, the process stops. Outputs "Drama" and "Crime" are predicted for input x_{12}. In summary, outputs "Romance" and "Crime" are equaly related to feature "Actor = T. Hanks". Therefore, it may be difficult to distinguish these two outputs based solely on this feature. However, if we are confident that a movie starred by "T. Hanks" should be classified as "Drama", then it is more likely that this movie should be classified as "Crime", rather than "Romance".

5.2 Empirical Results

In this section we will present the experimental results for the evaluation of the proposed multi-label classification algorithms, namely: LAC-MR-IO and LAC-MR-CO.

Setup In all the experiments we used ten-fold cross-validation and the final results of each experiment represent the average of the ten runs. All the results to be presented were found statistically significant based on a t-test at 95% confidence level.

Computational Environment The experiments were performed on a Linux-based PC with a Intel Pentium III 1.0 GHz processor and 1 GB RAM.

Evaluation Criteria The evaluation of multi-label classification algorithms is much more complicated than the evaluation of single-label ones. We used three evaluation criteria that were proposed in [4]:

- Hamming loss (h_{x_i}) : Evaluates how many times input x_i is misclassified (i.e., an output not related to x_i is predicted or an output related to x_i is not predicted), as show in Eq.5.1, where p is the number of possible outputs and Δ stands for the symmetric difference between the set of predicted outputs (P_{x_i}) and the set of true outputs (A_{x_i}) for input x_i.

$$h_{x_i} = \frac{1}{p} \mid P_{x_i} \Delta A_{x_i} \mid . \tag{5.1}$$

- Ranking loss (r_{x_i}): Evaluates the average fraction of output pairs $(c_j, c_k$, for which $c_k \in A_{x_i}$ and $c_j \notin A_{x_i})$ that are reversely ordered (i.e., $j > k$), as shown in Eq. 5.2 (where $\overline{A_{x_i}}$ denotes the complementary set of A_{x_i}).

$$r_{x_i} = \frac{\left|\left\{(c_k, c_j) \in A_{x_i} \times \overline{A_{x_i}} : j > k\right\}\right|}{|A_{x_i}|\,|\overline{A_{x_i}}|}. \tag{5.2}$$

- One-error (o_{x_i}): Evaluates the label ranking performance from a restrictive perspective as it only determines if the top-ranked label is present in the set of proper outputs (A_{x_i}) of input x_i, as shown in Eq. 5.3.

$$o_{x_i} = \begin{cases} 0 & \text{if most likely output is in} A_{x_i} \\ 1 & \text{otherwise} \end{cases} \tag{5.3}$$

The overall classification performance is obtained by averaging each criterion, that is:

$$\text{Hamming loss} = \frac{1}{|T|} \times \sum_{x_i \in T} h_{x_i} \tag{5.4}$$

$$\text{Ranking loss} = \frac{1}{|T|} \times \sum_{x_i \in T} r_{x_i} \tag{5.5}$$

$$\text{One error} = \frac{1}{|T|} \times \sum_{x_i \in T} o_{x_i} \tag{5.6}$$

5.2.1 Document Categorization

Two collections were used in the experiments. The first collection, which is called ACM-DL (first level), was extracted from the first level of the ACM Computing Classification System (http://www.portal.acm.org/dl.cfm), comprising a set of 81,251 documents labeled using the 11 first level categories of ACM. The second collection, ACM-DL (second level) contains the same set of documents of ACM-DL (first level), but these documents are labeled using the 81 second level categories. In both collections, each document is described by its title and abstract, citations, and authorship, resulting in a huge and sparse feature space. For ACM-DL (first level), the average number of labels (or outputs) for each document is 2.55, while for ACM-DL (second level) the average number of labels for each document is 2.82.

Baselines The evaluation is based on a comparison involving ML-SVM [2].

Parameters For ML-SVM, polynomial kernels of degree 8 were used. For LAC-MR-IO and LAC-MR-CO, σ_{min} was set to 0.01. For LAC-MR-IO, Δ_{min} was set to 0.25.

Analysis Table 5.2 shows categorization performance for different classification algorithms. Best results, including statistical ties, are shown in bold. ML-SVM and LAC-MR-IO shown competitive performance, and LAC-MR-CO is the best

Table 5.2 Categorization performance for different algorithms

Algorithms	First level			Second level		
	Hamm. loss	Rank. loss	One-error	Hamm. loss	Rank. loss	One-error
ML-SVM	0.225	0.194	**0.244**	0.327	0.299	0.348
LAC-MR-IO	0.222	0.216	**0.238**	0.319	0.294	**0.331**
LAC-MR-CO	**0.187**	**0.179**	**0.238**	**0.285**	**0.273**	**0.331**

performer. To verify if the association between labels was properly explored by LAC-MR-CO, we checked whether the explicitly correlated categories shown in the ACM Computing Classification System (http://www.acm.org/class/1998/overview.html) were indeed used. We verified that some of these explicitly correlated categories often appear together in the predicted label combination (i.e., *Files* and *DatabaseManagement*, or *Simulation/Modeling* and *Probability/Statistics*). We further verified that some of the associated labels appear more frequently in the predictions performed by LAC-MR-CO than it was observed in the predictions performed by the other algorithms.

5.2.2 Gene Functional Analysis

Genes play a fundamental role in life. Thus, predicting the function of a certain gene is of great interest. In this section we evaluate LAC-MR-IO and LAC-MR-CO for sake of predicting the gene functional classes of the *Yeast Saccharomyces cerevisiae*, which is one of the best studied organisms. More specifically, the YEAST dataset [4] is investigated. The whole set of functional classes is structured into hierarchies up to 4 levels deep 4. In our evaluation, only functional classes in the top hierarchy are considered. The dataset is composed of a set of 2,417 genes. Each gene is described by the concatenation of micro-array expression data and phylogenetic profile, and is associated with a set of functional classes. There are 14 possible class labels (functions), and the average number of labels for each gene is 4.24.

Baselines The evaluation is based on a comparison involving BoosTexter [4], ADTBoost.MH [1], and ML-SVM [2]. We believe that these methods are representative of some of the most effective multi-label methods available.

Parameters For BoosTexter and ADTBoost.MH, the number of boosting rounds was set to 500 and 50, respectively. For ML-SVM, polynomial kernels of degree 10 were used. For LAC-MR-IO and LAC-MR-CO, σ_{min} was set to 0.01. For LAC-MR-IO, Δ_{min} was set to 0.25.

Analysis Table 5.3 shows the results. Best results, including statistical ties, are shown in bold. The YEAST dataset is considered complex, with strong dependencies among labels. LAC-MR-CO provide gains of 24% in terms of one-error, considering BoosTexter as the baseline. The reason is that the simple decision function used by BoosTexter is not suitable for this complex dataset.

Table 5.3 Categorization performance for different algorithms

Algorithms	Hamming loss	Ranking loss	One-error
BoosTexter	0.220	0.186	0.278
ADTBoost.MH	0.207	–	0.244
ML-SVM	0.196	0.163	0.217
LAC-MR-IO	0.191	0.164	**0.213**
LAC-MR-CO	**0.179**	**0.150**	**0.213**

Also, LAC-MR-IO and LAC-MR-CO are able to explore many more associations than ADTBoost.MH. LAC-MR-CO performs much better than ML-SVM since it is able to explore dependencies between labels.

References

1. Comité, F., Gilleron, R., Tommasi, M.: Learning multi-label alternating decision trees from texts and data. In: Proceedings of the International Conference on Machine Learning and Data Mining in Pattern Recognition (MLDM), pp. 35–49. Springer (2003)
2. Elisseeff, A., Weston, J.: A kernel method for multi-labelled classification. In: Proceedings of the Annual Conference on Neural Inf. Processing Systems (NIPS), pp. 681–687. MIT Press, Cambridge (2001)
3. Menezes, G., Almeida, J., Belém, F., Gonçalves, M., Lacerda, A., de Moura, E.S., Pappa, G., Veloso, A., Ziviani, N.: Demand-driven tag recommendation. In: Proceedings of the European Conference on Principles of Data Mining and Knowledge Discovery (ECML/PKDD), pp. 402–417 (2010)
4. Schapire, R., Singer, Y.: Boostexter: A boosting-based system for text categorization. Mach. Learn. **39**(2–3), 135–168 (2000)
5. Veloso, A., Meira, W., Jr, Gonçalves, M., Zaki, M., Multi-label lazy associative classification. In: Proceedings of the European Conference on Principles of Data Mining and Knowledge Discovery (PKDD), pp. 605–612. Springer (2007)

Chapter 6
Competence–Conscious Associative Classification

Abstract The classification performance of an associative classification algorithm is strongly dependent on the statistic measure or metric that is used to quantify the strength of the association between features and classes (i.e., confidence, correlation, etc.). Previous studies have shown that classification algorithms produced using different metrics may predict conflicting outputs for the same input, and that the best metric to use is data-dependent and rarely known while designing the algorithm (Veloso et al. Competence–conscious associative classification. Stati Anal Data Min 2(5–6):361–377,2009; The metric dillema: competence–conscious associative classification. In: Proceeding of the SIAM Data Mining Conference (SDM). SIAM, 2009). This uncertainty concerning the optimal match between metrics and problems is a dilemma, and prevents associative classification algorithms to achieve their maximal performance. A possible solution to this dilemma is to exploit the competence, expertise, or assertiveness of classification algorithms produced using different metrics. The basic idea is that each of these algorithms has a specific sub-domain for which it is most competent (i.e., there is a set of inputs for which this algorithm consistently provides more accurate predictions than algorithms produced using other metrics). Particularly, we investigate stacking-based meta-learning methods, which use the training data to find the domain of competence of associative classification algorithms produced using different metrics. The result is a set of competing algorithms that are produced using different metrics. The ability to detect which of these algorithms is the most competent one for a given input leads to new algorithms, which are denoted as *competence–conscious associative classification algorithms*.

keywords Competence · Association metrics · Meta-learning · Stacking · Document categorization · Spam detection

A. Veloso and W. Meira Jr., *Demand-Driven Associative Classification*,
SpringerBriefs in Computer Science, DOI: 10.1007/978-0-85729-525-5_6,

6.1 Algorithms for Competence–Conscious Associative Classification

Next we present three algorithms for multi-metric associative classification. The first one, which will be referred to as LAC-MR-SD (standing for LAC-MR with self-delegation), delegates the metric to be used for producing $f_S^{x_i}$. The second one, which will be referred to as LAC-MR-OC (standing for LAC-MR with output-centric metric selection), groups the competence of metrics according to the outputs. The last one, which will be referred to as LAC-MR-IC (standing for LAC-MR with input-centric metric selection), is much finer-grained and associates the competence of metrics to inputs.

6.1.1 Metrics

Next, we present several metrics for measuring the strength of association between a set of features (X) and classes $(c_1, c_2, \ldots, c_p)$. Some of these metrics are popular ones [1, 13], while others were recently used in the context of associative classification [3]. These metrics interpret association using different definitions. We believe that these definitions are different enough to indicate that the corresponding algorithms may present some diversity.

- Confidence (γ_1) [1]: This metric was defined in Eq. 3.1. Its value ranges from 0 to 1.
- Added Value (γ_2) [9]: This metric measures the gain in accuracy obtained by using rule $X \rightarrow c_j$ instead of always predicting c_j, as shown in Eq. 6.1. Negative values indicate that always predicting c_j is better than using the rule. Its value ranges from -1 to 1.

$$\gamma_2 = p(c_j|X) - p(c_j) \qquad (6.1)$$

- Certainty (γ_3) [10]: This metric measures the increase in accuracy between rule $X \rightarrow c_j$ and always predicting c_j, as shown in Eq. 6.2. It assumes values smaller than 1.

$$\gamma_3 = \frac{p(c_j|X) - p(c_j)}{p(\overline{c_j})} \qquad (6.2)$$

- Yules'Q (γ_4) and Yules'Y (γ_5) [13]: These metrics are based on odds value, as shown in Eqs. 6.3 and 6.4, respectively. Their values range from -1 to 1. The value 1 implies perfect positive association between X and c_j, value 0 implies no association, and value -1 implies perfect negative association.

$$\gamma_4 = \frac{p(X \cup c_j) \times p(\overline{X} \cup \overline{c_j}) - p(X \cup \overline{c_j}) \times p(\overline{X} \cup c_j)}{p(X \cup c_j) \times p(\overline{X} \cup \overline{c_j}) + p(X \cup \overline{c_j}) \times p(\overline{X} \cup c_j)} \qquad (6.3)$$

$$\gamma_5 = \frac{\sqrt{p(X \cup c_j) \times p(\overline{X} \cup \overline{c_j})} - \sqrt{p(X \cup \overline{c_j}) \times p(\overline{X} \cup c_j)}}{\sqrt{p(X \cup c_j) \times p(\overline{X} \cup \overline{c_j})} + \sqrt{p(X \cup \overline{c_j}) \times p(\overline{X} \cup c_j)}} \qquad (6.4)$$

- Strength score (γ_6) [3]: This metric measures the correlation between X and c_j, but it also takes into account how X is correlated to the complement of c_j (i.e., $\overline{c_j}$), as shown in Eq. 6.5. Its value ranges from 0 to ∞.

$$\gamma_6 = \frac{p(X|c_j) \times p(c_j|X)}{p(X|\overline{c_j})} \qquad (6.5)$$

- Support (γ_7) [1]: This metric was defined in Eq. 3.2. Its value ranges from 0 to 1.
- Weighted Relative Confidence (γ_8) [10]: This metric trades off accuracy and generality, as shown in Eq. 6.6. The first component is the accuracy gain that is obtained by using rule $X \rightarrow c_j$ instead of always predicting c_j. The second component incorporates generality.

$$\gamma_8 = (p(c_j|X) - p(c_j)) \times p(X) \qquad (6.6)$$

Although we focus our analysis only on these metrics, the algorithms to be introduced are general and able to exploit any number of metrics, transparently.

Algorithm LAC-MR-SD Selecting an appropriate metric is a major issue while designing an associative classification algorithm. Algorithms produced by different metrics often present different classification performance. Depending on the characteristics of the problem, some metrics may be more suitable than others. Given a set composed of algorithms, $C_{\gamma_1}, C_{\gamma_2}, \ldots, C_{\gamma_q}$, which were produced using different metrics, we must select which algorithm is the one most likely to perform a correct prediction. Equation 3.5 can be used to estimate the reliability of a prediction, and this information can be used to select the most reliable prediction performed, considering all constituent classification algorithms. This is the approach used by LAC-MR-SD, which is illustrated in Algorithm 10. For a given input x_i, the predicted output is the one which is associated with the highest likelihood $\hat{p}(c_j|x_i)$ amongst all competing algorithms. The basic idea is to use the most reliable prediction (among the predictions performed by all competing algorithms) to select the output for x_i.

Although simple, LAC-MR-SD does not exploit the competence of each constituent algorithm. In fact, each base algorithm simply decides by itself the inputs for which it will predict the output, not meaning that the selected inputs belong to its domain of competence.

Algorithm 10 Finding $f_S^{x_i}$, according to LAC-MR-SD.

Require: The training data S, and an input $x_i \in T$
Ensure: $f_S^{x_i}$

1: $S^{x_i} \Leftarrow S$ projected according to x_i
2: $R^{x_i} \Leftarrow$ rules $X \rightarrow c_j$ extracted from S^{x_i}
3: **for** each competing algorithm C_{γ_q} **do**
4:     produce candidate functions $c_{\gamma_q S}^{x_i}$ using rules in R^{x_i}
5: **end for**
6: return $f_S^{x_i}$ which is the function that provides the highest likelihood $\hat{p}(c_j|x_i)$, amongst all candidate functions $c_{\gamma_q S}^{x_i}$

6.1.2 Domain of Competence

The optimal match between metrics and problems is valuable information. In this section we present an approach to estimate such matching. The proposed approach may be viewed as an application of Wolpert's stacked generalization [15]. From a general point of view, stacking can be considered a meta-learning method, as it refers to the induction of algorithms over inputs that are, in turn, the predictions of other algorithms induced from the training data.

Algorithm 11 Enhancing the training data with the competence of each competing algorithm.

Require: The original training data S, and a cross-validation parameter k
Ensure: The enhanced training data S_e

1: split S into k partitions, so that $S=\{d_1 \cup d_2 \cup \ldots \cup d_k\}$
2: $S_e \Leftarrow \emptyset$
3: **for** each partition d_t **do**
4:    **for** each input $x_i \in d_t$ **do**
5:      $\gamma \Leftarrow \emptyset$
6:      $\{S - d_t\}^{x_i} \Leftarrow \{S - d_t\}$ projected according to x_i
7:      $R^{x_i} \Leftarrow$ rules $X \rightarrow c_j$ extracted from $\{S\text{-}d\}^{x_i}$
8:      build different algorithms, $C_{\gamma_1}^t, C_{\gamma_2}^t, \ldots, C_{\gamma_q}^t$, using rules in R^{x_i}
9:      **for** each algorithm C_{γ_j} **do**
10:        **if** C_{γ_j} correctly predicts the class for x_i **then**
11:          $\gamma \Leftarrow \gamma \cup \gamma_j$
12:        **end if**
13:      **end for**
14:      $S_e \Leftarrow S_e \cup \{(x_i, y_i) \cup \gamma\}$
15:    **end for**
16: **end for**

The process starts by enhancing the original training data using the outputs predicted by the base algorithms, $C_{\gamma_1}, C_{\gamma_2}, \ldots, C_{\gamma_q}$. Algorithm 11 shows the basic steps involved in the process. Initially, the enhanced training data, S_e is empty. An example x_i, along with the competence of each algorithm with regard to x_i (i.e., which competing algorithm correctly predicted the output for x_i), is inserted into S_e. The process continues until all examples are processed. In the end, for each example $x_i \in S_e$ we have a list of competing algorithms that predicted the correct output for x_i, and this information enables learning the domains of competence of each algorithm.

Table 6.1 Training data given as an example of multimetric problem

Id	Output	Input $a_1\ a_2, \ldots,\ a_l$
1	c_1	1 3,..., 6
2	c_1	1 3,..., 7
3	c_1	2 4,..., 6
4	c_2	2 4,..., 7
5	c_2	2 5,..., 8
6	c_2	2 4,..., 6
7	c_3	1 3,..., 9
8	c_3	2 5,..., 9
9	c_3	2 4,..., 8
10	c_3	2 4,..., 9

Table 6.2 Enhanced training data, S_e

Id	Output	Input $a_1\ a_2, \ldots,\ a_l$	Competent Metric (s) (per instance)	Most competent Metric (s) (per class)
1	c_1	1 3,..., 6	γ_2	
2	c_1	1 3,..., 7	$\gamma_1\gamma_3$	γ_1
3	c_1	2 4,..., 6	γ_1	
4	c_2	2 4,..., 7	$\gamma_1\gamma_2$	
5	c_2	2 5,..., 8	$\gamma_1\gamma_2\gamma_3$	γ_1
6	c_2	2 4,..., 6	γ_1	
7	c_3	1 3,..., 9	γ_2	
8	c_3	2 5,..., 9	$\gamma_2\gamma_3$	γ_2
9	c_3	2 4,..., 8	$\gamma_1\gamma_2\gamma_3$	
10	c_3	2 4,..., 9	γ_2	

Example To illustrate the process, consider the example shown in Tables 6.1 and 6.2. Table 6.1 shows the original training data, S. Using the process described in Algorithm 11, the competence of each algorithm with regard to each input is appended to S, resulting in the enhanced training data, S_e, which is shown in Table 6.2. In this case, for a given example x_i, metric γ_j is shown if the corresponding algorithm C_{γ_j} has correctly predicted the output for input x_i using the stacking procedure. The enhanced training data, S_e, can be exploited in several ways. In particular, we will use S_e to produce competence–conscious associative classification algorithms, as it will be discussed next.

6.1.3 Competence–Conscious Metric Selection

In the following we present algorithms that exploit S_e to produce functions $f_S^{x_i}$. The challenge, in this case, is to properly select a competent metric for a specific input. The competence–conscious algorithms to be presented differ in how they perform the analysis of the domains of competence of the competing algorithms.

Algorithm LAC-MR-OC The competence of algorithms produced using different metrics are often associated with certain outputs (or classes). Some metrics, for instance, produce algorithms which show preference for more frequent classes, while others produce algorithms which show preference for less frequent ones. As an illustrative example, please consider Table 6.2. Algorithm derived from metric γ_1 is extremely competent for inputs that are related to outputs c_1 and c_2. On the other hand, if we consider inputs that are related to c_3, the algorithm derived from metric γ_2 perfectly classifies all inputs. This information (which is shown in the last column of Table 6.2) may be used to produce output-centric competence–conscious algorithms. The process is depicted in Algorithm 12. It starts with a meta-classifier, M, which learns the most competent base algorithm for a given class. Specifically, instead of extracting rules $X \rightarrow c_j$, the meta-classifier extracts rules $X \rightarrow \gamma_i$, which maps features (i.e., in the third column of Table 6.2) to metrics (i.e., in the fifth column of Table 6.2). Then, for each input $x_i \in T$, the meta-classifier indicates the most competent base algorithm, C_{γ_j}, that is then used to produce $f_S^{x_i}$.

Algorithm 12 Finding $f_S^{x_i}$, according to LAC-MR-OC.

Require: The enhanced training data S_e (i.e., the 3rd and 5th columns of Table 6.2), and an input $x_i \in T$

Ensure: $f_S^{x_i}$

1: $S_e^{x_i} \Leftarrow S_e$ projected according to x_i
2: **for each** metric γ_t **do**
3:     $R_{\gamma_t}^{x_i} \Leftarrow$ rules $X \rightarrow \gamma_t$ extracted from $S_e^{x_i}$
4:     estimate $\hat{p}(\gamma_t | x_i)$, according to Equation 3.5
5: **end for**
6: let C_{γ_j}, such that $\hat{p}(\gamma_j | x_i) \geq \hat{p}(\gamma_t | x_i) \forall t \neq j$, be the most competent algorithm for input x_i
7: **return** $f_S^{x_i}$ which is produced by C_{γ_j}

Algorithm LAC-MR-IC Although the competence of some base algorithms are associated with certain classes, specific inputs may be better classified using other base algorithms. In such cases, a finer-grained analysis of competence is desired. As an illustrative example, consider again Table 6.2. Although algorithm derived from metric γ_1 is the most competent one to predict the outputs for inputs that are related to class c_1, algorithm derived from metric γ_2 is the only one which competently classifies input 1 (which is related to c_1). Again, a meta-classifier, M, is used to explore such cases. The process is depicted in Algorithm 13. In this case, the meta-classifier learns the most competent metric by extracting rules of the form $X \rightarrow \gamma_j$, which maps features (i.e., in the third column of Table 6.2) to metrics (i.e., in the fourth column of Table 6.2). Then, for each input $x_i \in T$, the meta-classifier indicates the most competent base algorithm, C_{γ_j}, that is then used to produce $f_S^{x_i}$.

The main advantage of LAC-MR-OC and LAC-MR-IC is that, in practice, multiple metrics produce competent algorithms for a particular input x_i, but M

needs to predict only one of them (competent algorithms are not mutually exclusive, and thus, in practice, multiple metrics produce competent algorithms for x_i). This redundancy in competence that exists when different metrics are taken into account, may increase the chance of selecting a competent algorithm.

Algorithm 13 Finding $f_S^{x_i}$, according to LAC-MR-IC.

Require: The enhanced training data S_e (i.e., the 3rd and 4th columns of Table 6.2), and an input
   $x_i \in T$
Ensure: $f_S^{x_i}$

1: $S_e^{x_i} \Leftarrow S_e$ projected according to x_i
2: **for** each metric γ_t **do**
3:    $R_{\gamma_t}^{x_i} \Leftarrow$ rules $X \rightarrow \gamma_t$ extracted from $S_e^{x_i}$
4:    estimate $\hat{p}(\gamma_t|x_i)$, according to Equation 3.5
5: **end for**
6: let C_{γ_j} such that $\hat{p}(\gamma_j|x_i) \geq \hat{p}(\gamma_t|x_i) \forall t \neq j$, be the most competent algorithm for input x_i
7: **return** $f_S^{x_i}$ which is produced by C_{γ_j}

6.2 Empirical Results

In this section we will present the empirical results for the evaluation of the proposed multi-metric classification algorithms, which include LAC-MR-SD, LAC-MR-OC, and LAC-MR-IC.

Setup In all the experiments we used ten-fold cross-validation and the final results of each experiment represent the average of the ten runs. All the results to be presented were found statistically significant based on a t-test at 95% confidence level.

Baselines The evaluation is based on a comparison involving SVM algorithms and against ER (standing for External Referee), which is a combination method proposed in [11] (in this case, the competing algorithms are $C_{\gamma_1}, \ldots, C_{\gamma_8}$, but the most competent algorithm for each input is selected using a decision tree referee). For SVM, we used the implementation available at http://svmlight.joachims.org (version 3.0). We used our own implementation of ER.

Computational Environment The experiments were performed on a Linux-based PC with a Intel Pentium III 1.8 GHz processor and 1 GB RAM.

Bounds for multi-metric associative classification We derived simple lower and upper bounds for the classification performance of LAC-MR-SD, LAC-MR-OC, and LAC-MR-IC. The lower bound is the performance that is obtained by randomly selecting a competent algorithm. Clearly, this lower bound increases with the redundancy between the base algorithms (this redundancy exists because competent algorithms are not mutually exclusive, and, thus, for a particular input x_i, multiple base algorithms may be competent). The upper bound is the classification performance that would be obtained by an oracle which always predicts a competent base algorithm (note that perfect performance is not always possible, since it may not exist a competent algorithm for some inputs). Clearly, this upper bound increases with the accuracy and diversity associated with base algorithms.

6.2.1 Document Categorization

In this section we will evaluate the proposed algorithms using a collection of documents extracted from the ACM digital library. This is the same collection used in the experiments shown in Sect. 3.3.2. There 6,682 documents, which were labeled under 8 first level categories of ACM, namely: Hardware (C1), Computer Systems Organization (C2), Software (C3), Computing Methodologies (C4), Mathematics of Computing (C5), Information Systems (C6), Theory of Computation (C7), Computing Milieux (C8).

Parameters As suggested by the grid parameter search tool in LibSVM [5], polynomial kernel of degree 6 was used in the experiments.

Evaluation Criteria Categorization performance for the various methods being evaluated, is expressed through $MicF_1$.

Analysis Using the rules extracted from ACM-DL, we can analyze the relationship between the widely used confidence metric (γ_1) with other metrics, as shown in Fig. 6.1 (to ease the observation of this relationship, we also include, in each graph, a thicker line which indicates the corresponding confidence value). Each point in the graphs corresponds to a rule, for which it is shown the values of some metrics (i.e., confidence in the x-axis and another metric in the y-axis). Different lines are associated with different outputs (i.e., classes). Clearly, each metric has its particular behavior with varying values of confidence. We will use these relationships to understand some of the results to be presented. For lower values of confidence, Added Value (γ_2) has a preference for less frequent classes, but, after a certain confidence value, the preference is for more frequent classes. Certainty (γ_3) always prefer less frequent classes, but linearly approaches confidence as its value increases. Yules'Q (γ_4) and Yules'Y (γ_5) have a similar behavior, showing preference for less frequent classes and hardly penalizing associations with low confidence values. Strength Score (γ_6) and Weighted Relative Confidence (γ_8) both prefer less frequent classes, but Strength score shows a non-proportional preference for associations with higher values of confidence. The relationship between confidence and support (γ_7) is omitted, but, by definition, support shows a preference for more frequent classes.

Table 6.3 shows the classification performance obtained by different base algorithms. Best results, including statistical ties, are shown in bold. We will first analyze the performance associated with each category, and then the final classification performance, which is shown in the last line of the table. Algorithms produced by confidence (C_{γ_1}) and support (C_{γ_7}) performed very well in the most frequent categories (Software, Information Systems and Theory of Computer Science). On the other hand, inputs belonging to less frequent categories (Computer Methodologies, Mathematics of Computer Science, and Computer Science

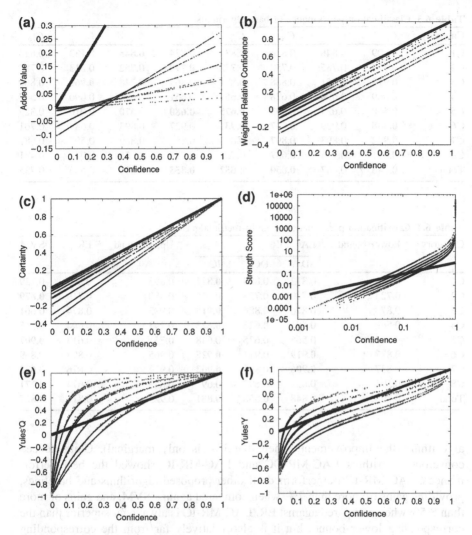

Fig. 6.1 Relationship between confidence and other metrics

Organization) were better classified using algorithms produced by Yules'Q (C_{γ_4}) and Yules'Y (C_{γ_5}). This is expected, and is in agreement with the behaviors depicted in Fig. 6.1 (algorithms produced by Yules'Y and Yules'Q show a preference for less frequent categories). The best base algorithm is the one that better balances its performance over all categories. Although the C_{γ_5} algorithm was not the best one for any specific category of ACM-DL, it was the best overall base algorithm.

Table 6.4 shows classification performance for multi-metric algorithms. Best results, including statistical ties, are shown in bold. LAC-MR-SD shows a performance that is similar to the performance obtained by most of the base

Table 6.3 Classification performance of base algorithms

Category	C_{γ_1}	C_{γ_2}	C_{γ_3}	C_{γ_4}	C_{γ_5}	C_{γ_6}	C_{γ_7}	C_{γ_8}
C1	0.809	**0.846**	0.826	0.834	0.834	**0.848**	0.183	0.628
C2	0.714	0.785	0.758	0.772	**0.799**	0.752	0.313	0.785
C3	0.912	0.851	0.888	0.871	0.864	0.748	**0.960**	0.880
C4	0.569	**0.690**	0.628	0.657	0.661	0.676	0.090	0.547
C5	0.548	0.624	0.593	**0.675**	**0.680**	**0.670**	0.010	0.329
C6	**0.948**	0.929	**0.937**	**0.931**	0.927	0.893	0.689	0.761
C7	**0.922**	0.893	0.897	0.890	0.887	0.889	0.507	0.687
C8	0.641	0.715	0.687	0.721	0.729	**0.755**	0.071	0.481
Total	0.843	0.847	**0.850**	**0.852**	**0.855**	0.810	0.566	0.735

Table 6.4 Classification performance of multi-metric algorithms

Category	Lower bound	LAC-MR			Upper bound	ER	SVM
		SD	OC	IC			
C1	0.715	0.813	0.809	**0.821**	0.893	0.801	0.729
C2	0.723	0.730	0.738	0.766	0.880	0.719	**0.879**
C3	0.870	0.876	0.884	**0.918**	0.983	0.874	0.661
C4	0.562	0.581	**0.623**	**0.623**	0.795	0.604	0.515
C5	0.563	0.568	0.625	0.648	0.751	0.613	**0.907**
C6	0.877	**0.919**	0.911	**0.925**	0.965	0.898	0.869
C7	0.837	**0.906**	0.895	**0.902**	0.922	0.876	0.672
C8	0.591	0.654	0.697	0.697	0.823	0.674	**0.771**
Total	0.798	0.848	0.858	**0.881**	0.925	0.811	0.827

algorithms (the improvement, when it exists, is only marginal). Competence–conscious algorithms LAC-MR-OC and LAC-MR-IC showed the best performances. LAC-MR-IC outperformed all other proposed algorithms and baselines, providing gains of more than 7%, when compared against SVM, and gains of more than 8.5% when compared against ER. LAC-MR-IC is always far superior than the corresponding lower bound, but it is also relatively far from the corresponding upper bound.

We also performed an analysis on how the different base algorithms were used by LAC-MR-OC and LAC-MR-IC, as can be seen in Fig. 6.2. LAC-MR-OC utilized only few base algorithms, specially C_{γ_2}, C_{γ_3}, and C_{γ_7}. Metric γ_4 was used to produce algorithms to only one category, and metrics γ_5 and γ_8 were not used to produce any base algorithm at all (this is because the corresponding algorithms were not the most competent in any category of ACM, and therefore are not considered by LAC-MR-OC). LAC-MR-IC, on the other hand, utilized all base algorithms, specially C_{γ_1}, C_{γ_2} and C_{γ_3}. Both LAC-MR-OC and LAC-MR-IC make large utilization of base algorithms C_{γ_2} and C_{γ_3}. For LAC-MR-OC, some areas of expertise can be easily detected. Base algorithm C_{γ_2} is considered competent for categories Hardware and Computer Science Organization, while C_{γ_3} is considered

Fig. 6.2 Utilization of base algorithms

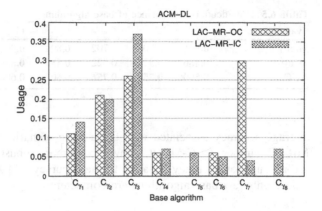

Fig. 6.3 Distribution of competent algorithms

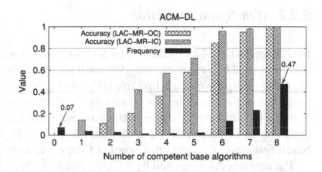

competent for category Information Systems. For LAC-MR-IC, areas of expertise are finer grained, but with manual inspection we detected that C_{γ_1} is considered competent for category Computer Science Organization, and C_{γ_3} is considered competent for category Milieux.

We finalize this set of experiments by analyzing one of the reasons of the good performance showed by LAC-MR-IC. Figure 6.3 shows the accuracy associated with scenarios for which a different number of base algorithms are competent. The frequency of occurrence of each scenario is also shown (note that both accuracy and frequency values are shown in the y-axis). As it can be seen, for more than 7% of the inputs in the test set, no base algorithm is competent, and, obviously, these inputs were misclassified (this means that the inclusion of other metrics may improve classification performance in this dataset). As expected, accuracy increases with the number of competent base algorithms. For almost half of the inputs in the test set all base algorithms were competent. In these scenarios, there is no risk of misclassification, since a base algorithm produced by any metric will perform a correct prediction. The accuracy associated with scenarios where only 7 and only 6 of the base algorithms are competent, is also extremely high (respectively, 99 and 96%). These three scenarios (i.e., 8, only 7, and only 6 base algorithms are simultaneously competent) correspond to 86% of the inputs, and the

Table 6.5 Classification performance of base algorithms

Evaluation target	C_{γ_1}	C_{γ_2}	C_{γ_3}	C_{γ_4}	C_{γ_5}	C_{γ_6}	C_{γ_7}	C_{γ_8}
MicF$_1$	**0.946**	0.704	0.702	0.894	0.901	**0.948**	**0.946**	0.880
MacF$_1$	0.486	0.522	0.522	0.584	**0.589**	**0.592**	0.486	**0.587**
AUC	0.500	**0.756**	**0.756**	0.607	0.606	0.562	0.500	0.629

average accuracy associated with these three scenarios is almost 98% for LAC-MR-IC. Further, LAC-MR-IC shows to be more robust than LAC-MR-OC, providing superior accuracy (relative to the accuracy of LAC-MR-OC) in scenarios where only few base algorithms are competent.

6.2.2 Web Spam Detection

In this application the objective is to detect malicious actions aimed at the ranking functions used by search engines. We used a dataset obtained from the Web Spam Challenge (http://webspam.lip6.fr/wiki/pmwiki.php). The dataset is very skewed (only 6% of the examples are spam pages). Each example is composed of direct features (i.e., number of pages in the host, number of characters in the host name, etc.) link-based features (i.e., in-degree, out-degree, PageRank, etc.) and content-based features (i.e., number of words in the page, average word length, etc.).

Parameters As suggested by the grid parameter search tool in LibSVM [5], we used a linear kernel with parameter C set to 5.00.

Evaluation Criteria For this application, classification performance is computed through MicF$_1$ and MacF$_1$ measures, and the area under the ROC curve [7].

Analysis Table 6.5 shows the classification performance obtained by different base algorithms. Best results, including statistical ties, are shown in bold. C_{γ_1} and C_{γ_7} showed impressive performance in terms of MicF$_1$. This is expected, because the vast majority of examples are legitimate pages, and confidence and support have preference for more frequent classes. On the other hand, C_{γ_1} and C_{γ_7} showed poor classification performance in terms of MacF$_1$ and AUC (i.e., no spam pages were detected). The remaining base algorithms were able to detect some spam pages, specially C_{γ_6}, which also shows impressive performance in terms of accuracy. In terms of AUC, C_{γ_2} and C_{γ_3} showed the best classification performance, amongst all base algorithms. Thus, different algorithms produced by different metrics show distinct performance depending on the evaluation target (i.e., MicF$_1$, MacF$_1$, or AUC).

Table 6.6 shows classification performance for multi-metric algorithms. Best results, including statistical ties, are shown in bold. LAC-MR-OC and LAC-MR-IC were the best performers in terms of MacF$_1$. Although LAC-MR-IC showed to be far from the optimal classification performance, it showed impressive gains when compared against SVM and ER, in terms of MacF$_1$ and AUC.

Table 6.6 Classification performance of multi-metric algorithms

Evaluation target	Lower bound	LAC-MR			Upper bound	ER	SVM
		SD	OC	IC			
MicF$_1$	0.852	0.861	0.870	0.897	0.990	0.866	**0.956**
MacF$_1$	0.588	0.594	0.609	**0.624**	0.947	0.586	0.504
AUC	0.662	0.730	0.718	**0.789**	0.908	0.725	0.512

References

1. Agrawal, R., Imielinski, T., Swami, A.: Mining association rules between sets of items in large databases. In: Proceedings of the International Conference on Management of Data (SIGMOD), pp. 207–216. ACM Press (1993)
2. Antonie, M., Zaïane, O., Holte, R.: Learning to use a learned model: a two-stage approach to classification. In: Proceedings of the International Conference on Data Mining (ICDM), pp. 33–42. IEEE Computer Society (2006)
3. Arunasalam, B., Chawla, S.: CCCS: a top-down associative classifier for imbalanced class distribution. In: Proceedings of the International Conference on Data Mining and Knowledge Discovery (KDD), pp. 517–522. ACM Press (2006)
4. Breiman, L.: Bagging predictors. Mach.Learn. **24**(2), 123–140 (1996)
5. Chang, C.-C., Lin, C.-J.: LIBSVM: a library for support vector machines, 2001. Available at http://www.csie.ntu.edu.tw/~cjlin/papers/libsvm.pdf
6. Ferri, C., Flach, P., Hernández-Orallo, J.: Delegating classifiers. In: Proceedings of the International Conference on Machine Learning (ICML), p. 37. ACM Press (2004)
7. Fürnkranz, J., Flach, P.: An analysis of rule evaluation metrics. In: Proceedings of the International Conference on Machine Learning (ICML), pp. 202–209. IEEE Computer Society (2003)
8. Gama, J., Brazdil, P.: Cascade generalization. Mach. Learn. **45**, 315–343 (2000)
9. Hilderman, R., Hamilton, H.: Evaluation of interestingness measures for ranking discovered knowledge. In: Proceedings of the Pacific-Asia Conference on Research and Development in Knowledge Discovery and Data Mining (PAKDD), pp. 247–259. Springer (2001)
10. Lavrac, N., Flach, P., Zupan, B.: Rule evaluation measures: a unifying view. Induct. Log. Prog. **1634**, 174–185 (1999)
11. Ortega, J., Koppel, M., Argamon, S.:Arbitrating among competing classifiers using learned referees. Knowl. Inf. Syst.**3**, 470–490 (2001)
12. Schapire, R.: A brief introduction to boosting. In: Proceedings of the International Joint Conference on Artificial Intelligence (IJCAI), pp. 1401–1406. Morgen Kaufmann, San Francisco (1999)
13. Tan, P., Kumar, V., Srivastava, J.: Selecting the right interestingness measure for association patterns. In: Proceedings of the International Conference on Data Mining and Knowledge Discovery (KDD), pp. 32–41. ACM Press (2002)
14. Tsymbal, A., Pechenizkiy, M., Cunningham, P.: Dynamic integration with random forests. In: Proceedings of the European Conference on Machine Learning (ECML), pp. 801–808. Springer (2006)
15. Wolpert, D.: Stacked generalization. Neural Netw. **5**(2), 241–259 (1992)

Chapter 7
Calibrated Associative Classification

Abstract Given an input x_i and an arbitrary output c_j, a classification algorithm usually works by estimating the probability of x_i being related to c_j (i.e., class membership probability). Well calibrated classification algorithms are those able to produce functions that provide accurate estimates of class membership probabilities, that is, the estimated probability $\hat{p}(c_j|x_i)$ is close to $p(c_j|\hat{p}(c_j|x_i))$, which is the true, (unknown) empirical probability of x_i being related to output c_j given that the probability estimated by the classification algorithm is $\hat{p}(c_j|x_i)$. Calibration is not a necessary property for producing an accurate approximation of the target function, and, thus, most of the research has focused on direct accuracy maximization strategies rather than on calibration. However, non-calibrated functions are problematic in applications where the reliability associated with a prediction must be taken into account (i.e., cost-sensitive classification, cautious classifications etc.). In these applications, a sensible use of the classification algorithm must be based on the reliability of its predictions (Veloso et al. Calibrated lazy associative classification, Inform Sci, 2009), and thus, the algorithm must produce well calibrated functions. In this chapter we introduce calibrated associative classification algorithms.

Keywords Calibrated probabilities · Empirical probability · Theoretical probability · Likelihood of membership · Entropy · Minimum description length · Document categorization · Revenue maximization

7.1 Algorithms for Calibrated Associative Classification

In this section we define calibrated algorithms, and then we propose methods to calibrate associative classification algorithms. Further, we present two algorithms that are calibrated using the proposed calibration methods. The first one, which

A. Veloso and W. Meira Jr., *Demand-Driven Associative Classification*,
SpringerBriefs in Computer Science, DOI: 10.1007/978-0-85729-525-5_7,
© Adriano Veloso 2011

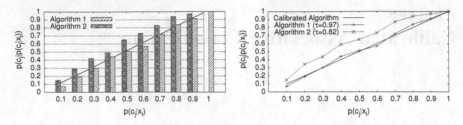

Fig. 7.1 Reliability diagram and τ-calibrated algorithms

will be referred to as LAC-MR-NC (standing for LAC-MR with naive calibration), is calibrated using a naive calibration method. The second one, which will be referred to as LAC-MR-EM (standing for LAC-MR with entropy minimization), is calibrated using a sophisticated calibration method based on entropy minimization.

7.1.1 τ-Calibrated Classification Algorithms

The calibration of a classification algorithm can be visualized using reliability diagrams. Diagrams for two arbitrary algorithms using an arbitrary dataset are depicted in Fig. 7.1 (left). These diagrams are built as follows [6]. First, the probability space (i.e., the x-axis) is divided into a number of bins, which was chosen to be 10 in our case. Probability estimates (i.e., theoretical probabilities) with value between 0 and 0.1 fall in the first bin, estimates with value between 0.1 and 0.2 fall in the second bin, and so on. The fraction of correct predictions associated with each bin, *which is the true, empirical probability* (i.e., $p(c|\hat{p}(c_j|x_i)))$, is plotted against the estimated probability (i.e., $\hat{p}(c_j|x_i)$). If the algorithm is well calibrated, the points will fall near the diagonal line, indicating that estimated probabilities are close to empirical probabilities. The degree of calibration of an algorithm, denoted as τ, is obtained by measuring the discrepancy between observed probabilities (o_i) and estimated probabilities (e_i), as shown in Eq. 7.1 (where k is the number of bins). Values of τ range from 0 to 1. A value of 0 means that there is no relationship between estimated probabilities and true probabilities. A value of 1 means that all points lie exactly on a straight line with no scatter. Algorithm 1 is better calibrated than Algorithm 2, as shown in Fig. 7.1 (Right).

$$\tau = 1 - \frac{1}{k}\sum_{i=1}^{k}\frac{(o_i - e_i)^2}{(o_i + e_i)^2} \tag{7.1}$$

Algorithm LAC-MR-NC To transform original probability estimates, $\hat{p}(c_j|x_i)$, into accurate well calibrated probabilities, we also use a method based on binning. The method starts by estimating membership probabilities using the training data, S. A typical method is ten-Fold Cross-Validation. In this case, S is divided into ten partitions, and at each trial, nine partitions are used for training, while the remaining partition is used to simulate a test set. After the ten trials, the algorithm will have stored in the set O, the membership probability estimates for all inputs in S. This process is shown in Algorithm 14.

Algorithm 14 Estimating membership probabilities.

Require: Examples in S
Ensure: For each input x_i in S, the corresponding membership probabilities $\hat{p}(c_1|x_i), \hat{p}(c_2|x_i), \ldots, \hat{p}(c_p|x_i)$, along with the correct output

1: $O \Leftarrow \emptyset$
2: Split S into 10 equal-sized partitions, $p_1, p_2, \ldots, p_{10}$
3: **for** each partition p_j **do**
4:    **for** each input $x_i \in p_j$ **do**
5:       Estimate probabilities, $\hat{p}(c_1|x_i), \hat{p}(c_2|x_i), \ldots, \hat{p}(c_p|x_i)$, using $\{S-p_i\}$ as training
6:       $O \Leftarrow O \cup \{(\hat{p}(c_1|x_i), v_1)\} \cup \ldots \cup \{(\hat{p}(c_p|x_i), v_p)\}$, where $v_j=1$ if c_j is the correct class for example x_i, and $v_j=0$ otherwise
7:    **end for**
8: **end for**
9: return O

Once the probabilities are estimated, a naive calibration method would proceed by first sorting these probabilities in ascending order (i.e., the probability space), and then dividing them into k equal-sized bins, each having pre-specified boundaries. An estimate is placed in a bin according to its value (i.e., values between 0 and $\frac{1}{k}$ are placed in the first bin, values between $\frac{1}{k}$ and $\frac{2}{k}$ in the second, and so on). The probability associated with a bin is given by the fraction of correct predictions that were placed in it. An estimate $\hat{p}(c_j|x_i)$ is finally calibrated by using the probability associated with the corresponding bin. Specifically, each bin $b_{l \mapsto u} \in B$ (with l and u being its boundaries) works as a map, relating estimates $\hat{p}(c_j|x_i)$ (such that $l \leq \hat{p}(c_j|x_i) < u$) to the corresponding calibrated estimates, $p_{b_{l \mapsto u}}$. Thus, this process essentially discretizes the probability space into intervals, so that the accuracy associated with the predictions in each interval is as reliable as possible.

Such naive method, however, may be disastrous as critical information may be lost due to in bin boundaries. Instead, we propose to use information entropy associated with candidate bins to select the boundaries [7].

Algorithm LAC-MR-EM This algorithm uses the information in O to initially find a threshold that minimizes the entropy over all possible partitions; and it is then recursively applied to both of the partitions induced by the threshold. To illustrate the method, suppose we are given a set of pairs $(\hat{p}(c_j|x_i), v)^1 \in O$. In this case, the entropy of O is given by Eq. 7.2.

---

[1] v can take the values 0 (the prediction is wrong) or 1 (otherwise), as shown in step 6 of Algorithm 14.

Fig. 7.2 Calculating bin boundaries for different categories (category "Data Mining" on the *left*, and category "Inf. Retrieval" on the *right*)

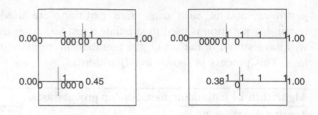

$$E(O) = -\frac{|(\hat{p}(c_j|x_i), 0) \in O|}{|O|} \times \log\left(\frac{|(\hat{p}(c_j|x_i), 0) \in O|}{|O|}\right)$$
$$-\frac{|(\hat{p}(c_j|x_i), 1) \in O|}{|O|} \times \log\left(\frac{|(\hat{p}(c_j|x_i), 1) \in O|}{|O|}\right) \qquad (7.2)$$

There is a threshold f, which is a boundary that induces two partitions of O ($b_{f\leq}$ and $b_{f>}$, where $b_{f\leq}$ contains pairs $(\hat{p}(c_j|x_i), v)$ for which $\hat{p}(c_j|x_i) \leq f$, and $b_{f>}$ contains pairs for which $\hat{p}(c_j|x_i) > f$). The selected threshold, t, is the one which minimizes the weighted average entropies, given by Eq. 7.3.

$$E(O,f) = \frac{|b_{f\leq}|}{|O|} \times E(b_{f\leq}) + \frac{|b_{f>}|}{|O|} \times E(b_{f>}) \qquad (7.3)$$

This method is then applied recursively to both of the partitions induced by t, $b_{t\leq}$ and $b_{t>}$, creating multiple intervals until a stopping criterion is fulfilled. Splitting stops if the information gain (the difference between the entropies before and after the split) is lower than the minimum description length [10] of the partition, and the final set of bins, B, is found. According to [7], the minimum description length induced by a threshold t over a partition O is the second term of the following inequality:

$$E(O) - E(O,t) > \frac{\log(|O| - 1)}{|O|} + \frac{\Delta(O,t)}{|O|} \qquad (7.4)$$

where $\Delta(O,t) = \log(3^k - 2) - (k \times E(O) - k_1 \times E(b_{t\leq}) - k_2 \times E(b_{t>}))$, and k_i is either 1 or 2 ($k_i = 1$ if the corresponding partition is pure, that is, if it contains only correct (or only incorrect) predictions; and $k_i = 2$ otherwise).

Finally, for each input x_i in the test set, $\hat{p}(c_1|x_i), \hat{p}(c_2|x_i), \ldots, \hat{p}(c_p|x_i)$ are estimated using S. Then, the estimated probabilities are calibrated using the accuracy associated with the appropriate bin in B, as shown in Algorithm 15.

Example Table 7.1 shows an illustrative example composed of ten documents extracted from a digital library. Each document belongs to one category. Such documents are given as training data. Several rules are extracted from these documents.

Table 7.1 Example using documents of a digital library

	Category	Features
d_1	Databases	Rules in Database Systems
d_2	Databases	Applications of Logic Databases
d_3	Databases	Hypertext Databases and Data Mining
d_4	Data Mining	Mining Association Rules in Large Databases
d_5	Data Mining	Database Mining: A Performance Perspective
d_6	Data Mining	Algorithms for Mining Association Rules
d_7	Inf. Retrieval	Text Databases and Information Retrieval
d_8	Inf. Retrieval	Information Filtering and Information Retrieval
d_9	Inf. Retrieval	Term Weighting Approaches in Text Retrieval
d_{10}	Inf. Retrieval	Performance of Information Retrieval Systems

---

Algorithm 15 Calibrating the probabilities.

Require: Examples in S, inputs in T, the calibrated probability $p_{b_{l \leftrightarrow u}}$ of each bin $b_{l \leftrightarrow u}$
Ensure: For each estimate $\hat{p}(c_j|x_i)$, the corresponding calibrated estimate $\hat{p}_c(c_j|x_i)$

1: **for** each input $x_i \in T$ **do**
2:     Estimate probabilities, $\hat{p}(c_1|x_i), \hat{p}(c_2|x_i), \ldots, \hat{p}(c_p|x_i)$, using S as training
3:     for each c_j do output $\hat{p}_c(c_j|x_i) = p_{b_{l \leftrightarrow u}}$, such that $l \leq \hat{p}(c_j|x_i) < u$
4: **end for**

Specifically, R^{d_1}, which is the set of rules (extracted from $\{S - p_1\}$) matching document d_1, includes:

1. text = system(s) $\rightarrow$ Inf. Retrieval($\theta = 1.00$)
2. text = {rule(s) $\wedge$ database(s)} $\rightarrow$ Data Mining($\theta = 1.00$)
3. text = rule(s) $\rightarrow$ Data Mining($\theta = 1.00$)
4. text = database(s) $\rightarrow$ Databases($\theta = 0.40$)
5. text = database(s) $\rightarrow$ Data Mining($\theta = 0.40$)
6. text = database(s) $\rightarrow$ Inf. Retrieval($\theta = 0.20$)

From such decision rules, class membership probabilities for document d_1 are estimated using Eq. 3.5, resulting in the following probabilities: $\hat{p}($Databases$|d_1) = 0.19$, $\hat{p}($Data Mining$|d_1) = 0.44$, and $\hat{p}($Inf. Retrieval$|d_1) = 0.37$. Membership probabilities for all documents are shown in Table 7.2, where the number between parenthesis indicates if the prediction is correct (1) or not (0).

Figure 7.2 (left) shows the process of setting bin boundaries by entropy minimization, for category "Data Mining". Initially, the probability space (which ranges from 0.00 to 1.00) is divided into two bins. The cut point at 0.45 gives an information gain which is higher than the minimum description length of initial bin. Now, there are two bins. The bin on the right (i.e., [0.45–1.00]) is not divided anymore, since additional cut points would not provide enough information gain. The bin on the left is further divided into two other bins. The cut point at 0.20 gives an information gain which is higher than the minimum description length of this bin. Then, the process stops because no more bins are created. Figure 7.2

Table 7.2 Class membership probabilities

	$\hat{p}(\text{Databases}|d)$	$\hat{p}(\text{Data Mining}|d)$	$\hat{p}(\text{Inf. Retrieval}|d)$
d_1	0.19 (1)	0.44 (0)	0.37 (0)
d_2	0.27 (1)	0.48 (0)	0.24 (0)
d_3	0.26 (1)	0.61 (0)	0.13 (0)
d_4	0.36 (0)	0.46 (1)	0.18 (0)
d_5	0.27 (0)	0.25 (1)	0.48 (0)
d_6	0.30 (0)	0.53 (1)	0.17 (0)
d_7	0.22 (0)	0.25 (0)	0.53 (1)
d_8	0.00 (0)	0.29 (0)	0.71 (1)
d_9	0.00 (0)	0.00 (0)	1.00 (1)
d_{10}	0.31 (0)	0.31(0)	0.38 (1)

Table 7.3 Bin boundaries and calibrated probabilities for each category

Databases		Data Mining		Inf. Retrieval	
Boundaries	Prob.	Boundaries	Prob.	Boundaries	Prob.
[0.00−0.17]	0.000	[0.00−0.20]	0.000	[0.00−0.38]	0.000
[0.17−1.00]	0.375	[0.20−0.45]	0.200	[0.38−0.52]	0.500
		[0.45−1.00]	0.500	[0.52−1.00]	1.000

(right) shows the same process of setting bin boundaries for category "Inf. Retrieval".

Bins obtained for each category are shown in Table 7.3. For this simplified example, only two bins are produced for category "Databases", and only three bins are produced for categories "Data Mining" and "Inf. Retrieval". The calibrated probability for each bin, which is the fraction of correct predictions within each bin, are also shown in Table 7.3.

7.2 Empirical Results

In this section we present the experimental results for the evaluation of the proposed calibrated algorithms, LAC-MR-NC (which is calibrated using the naive calibration method) and LAC-MR-EM (which is calibrated using the MDL-based entropy-minimization method).

Computational Environment The experiments were performed on a Linux-based PC with a Intel Pentium III 1.0 GHz processor and 1 GB RAM.

Baselines The evaluation is based on a comparison against current state-of-the-art calibrated algorithms, which include SVM [1], Naive Bayes [5], and Decision Tree classifiers [9]. After being calibrated using specific methods [2, 8, 11], these algorithms are respectively referred to as CaSVM, CaNB, and CaDT.

7.2.1 Document Categorization

For this application we used the ACM-DL dataset, which was described in Sect. 3.3.2. The classification algorithm must decide to which category a document belongs. However, the administrator of the digital library imposes an additional minimum accuracy requirement, acc_{min}, to the algorithm. In this case, the algorithm must estimate the total accuracy after each prediction is performed, and then it must decide to continue classifying documents (if the estimated accuracy is higher than acc_{min}) or to stop classification (if the estimated accuracy is lower than acc_{min}).

Setup In all the experiments we used ten-fold cross-validation and the final results of each experiment represent the average of the ten runs. All the results to be presented were found statistically significant based on a t-test at 95% confidence level.

Evaluation Criteria We used accuracy, τ, and the fraction of documents classified at acc_{min}, to assess classification performance.

Parameters For CaSVM we used linear kernels and set $C = 0.90$. These parameters were set according to the grid parameter search tool in LibSVM [3]. For CaNB and CaDT we used the default parameters, which were also used in other works [5]. For LAC-MR-NC, the number of bins was set to 5. For LAC-MR-NC and LAC-MR-EM, we set $\sigma_{min} = 0.001$.

Analysis The bins produced by different calibration methods are shown in Fig. 7.3. Bin boundaries are shown in the x-axis and the corresponding calibrated probabilities are shown in the y-axis. Coincidentally, the MDL-based Entropy-Minimization method also produced 5 bins, but with varying sizes.

After the bins were found, we apply LAC-MR to the test set, and we replace the original probability estimate (x-axis) by the calibrated probability associated with the corresponding bin (y-axis). The result of calibration is depicted in Fig. 7.4, which shows τ values for LAC-MR, before and after being calibrated with different methods (resulting in LAC-MR-NC and LAC-MR-EM algorithms). Other algorithms were also evaluated. The worst algorithm in terms of calibration is SVM with $\tau = 0.69$. After calibrating SVM, the corresponding algorithm, CaSVM, shows $\tau = 0.75$. NB and LAC-MR, with $\tau = 0.76$ and $\tau = 0.78$, respectively, are already better calibrated than CaSVM. These algorithms, when calibrated, show the best calibration degrees – CaNB with a $\tau = 0.91$, and LAC-MR-EM with $\tau = 0.97$. Next, we will evaluate how this difference in calibration affects the effectiveness of the algorithms.

We continue our analysis by evaluating each algorithm in terms of its ability for estimating the actual accuracy. Figure 7.5 shows the actual accuracy and the accuracy estimates obtained with each algorithm, so that the corresponding values can be directly compared.[2] As expected, LAC-MR-EM shows to be

---

[2] For each experiment, predictions were sorted from the most reliable to the least reliable.

Fig. 7.3 Bins produced for category "Information Systems"

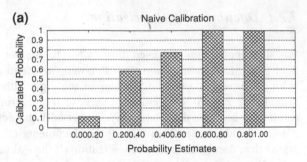

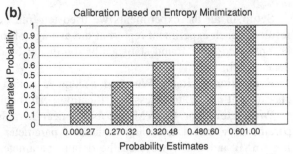

better calibrated than LAC-MR-NC. This is because the bins used by LAC-MR-NC are produced in an ad hoc way, while the bins used by LAC-MR-EM are produced following the entropy-minimization strategy. The direct consequence of applying such method is that a bin is likely to contain predictions which are as similar as possible. While in most of the cases CaNB and CaDT are well calibrated, CaSVM very often underestimates or overestimates the actual accuracy, and is thus poorly calibrated. The main reason of the poor performance of CaSVM is that Platt Scaling is prone to overfitting, since this calibration method is based on regression. The other calibration mechanisms apparently do not overfit as much. This explanation is supported by the results present in [4] (which show that Naive Bayes algorithms are much better calibrated than SVM algorithms).

If the administrator of the digital library specifies a threshold acc_{min} (i.e., the minimum acceptable accuracy of the algorithm), then the value of the algorithm resides in how many documents it is able to classify while respecting acc_{min}. Figure 7.5 shows the fraction of documents in the test set each algorithm is able to classify for a given value of acc_{min} (y-axis). Clearly, LAC-MR-EM is the best performer, except for acc_{min} values higher than 0.95, when the best performer is LAC-MR-NC. CaNB and CaDT are in close rivalry, with CaDT being slightly superior. In most of the cases, both CaNB and CaDT show to be superior than LAC-MR-NC. CaSVM, as expected, is the worst performer for all values of acc_{min}.

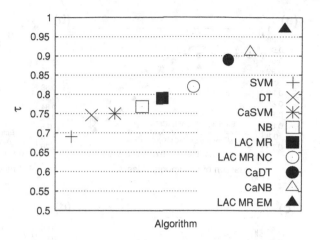

Fig. 7.4 Algorithms, before and after being calibrated

7.2.2 Revenue Maximization

For this application we used a dataset called KDD-98, which was used in KDD-CUP'98 contest. This dataset was provided by the Paralyzed Veterans of America (PVA), an organization which devises programs and services for US veterans. With a database of over 13 million donors, PVA is also one of the world's largest direct mail fund raisers.

The total cost invested in generating a request (including the mail cost), is $0.68 per piece mailed. Thus, PVA wants to maximize net revenue by soliciting only individuals that are likely to respond with a donation. The KDD-98 dataset contains information about individuals that have (or have not) made charitable donations in the past. The provided training data consists of 95,412 examples, and the provided test set consists of 96,367 inputs. Each example/input corresponds to an individual, and is composed of 479 features. The training data has an additional field that indicates the amount of the donation (a value $0 indicates that the individual have not made a donation). From the 96,367 individuals in the test set, only 4,872 are donors. If all individuals in the test set were solicited, the total profit would be only $10,547. On the other hand, if only those individuals that are donors were solicited, the total profit would be $72,764. Thus, the classifier must choose which individuals to solicit a new donation from. According to [11], the optimal net maximization strategy is to solicit an individual x_i if and only if $\hat{p}(\text{donate}|x_i) > \frac{0.68}{y(x_i)}$, where $y(x_i)$ is the expected amount donated by x_i.[3] Thus, in addition to calculating $\hat{p}(\text{donate}|x_i)$, the algorithm must also estimate $y(x_i)$.

Estimating the Donation Amount and $\hat{p}(\text{donate}|x)$ Each donation amount (i.e., $200, $199, ..., $1, and $0) is considered as an output. Thus, for an individual

---

[3] The basic idea is to solicit a person x_i for whom the expected return $\hat{p}(\text{donate}|x_i)y(x_i)$ is greater than the cost of mailing the solicitation.

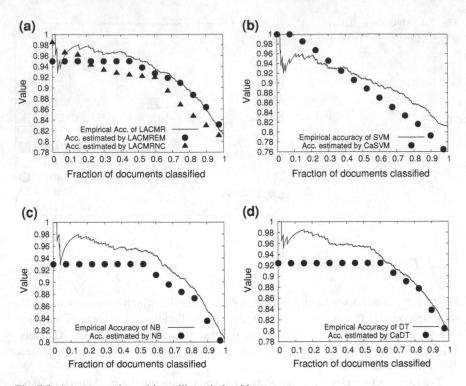

Fig. 7.5 Accuracy estimated by calibrated algorithms

x_i, rules of the form $X \rightarrow y$ (with $X \subseteq x_i$) are extracted from S, and Eq. 3.5 is used to estimate the likelihood of each amount (i.e., $\hat{p}(y = \$200|x_i), \hat{p}(y = \$199|x_i), \ldots, \hat{p}(y = \$0|x_i)$). The donation amount, $y(x_i)$, is finally estimated by a linear combination of the probabilities associated with each amount, as shown in Eq. 7.5. The probability of donation, $\hat{p}(donate|x_i)$, is simply given by $1 - \hat{p}(y = \$0|x_i)$.

$$y(x_i) = \sum_{i=\$0}^{\$200} i \times \hat{p}(y = i|x_i) \tag{7.5}$$

Parameters For CaSVM we used linear kernels and set $C = 2.00$. These parameters were set according to the grid parameter search tool in LibSVM [3]. For CaNB and CaDT we used the default parameters, which were also used in other works [5]. For LAC-MR-NC, we evaluate four different configurations, with 5, 8, 10, and 15 bins. For LAC-MR-NC and LAC-MR-EM, we set $\sigma_{min} = 0.001$.

Evaluation Criteria We used profit as the primary metric for assessing the effectiveness of the algorithms for net revenue optimization. For assessing the accuracy of probability estimates, we use the mean squared error (MSE). Calibration degree, τ, was also used.

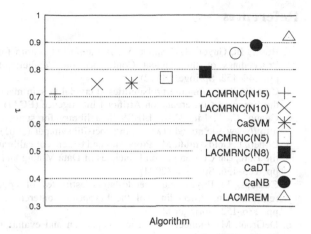

Fig. 7.6 Comparing calibration methods in terms of τ

Analysis The calibration degree, τ, achieved by each algorithm, is shown in Fig. 7.6. CaSVM, LAC-MR-NC(N15) and LAC-MR-NC(N10) achieved the lowest calibration degrees. This is because the naive calibration method over-fitted the training data (i.e., too many bins incur small-sized bins for which the corresponding accuracy may not be reliable, and for CaSVM, the Platt Scaling method is based on regression). LAC-MR-EM and CaNB are the best per-formers, achieving τ values as high as 0.94. Next we will analyze the effec-tiveness of algorithms with different calibration degrees for net revenue optimization.

Table 7.4 shows the effectiveness of each algorithm. In all cases, the differences in profit are much more accentuated than the differences in MSE. As it can be seen, LAC-MR achieved the lowest profit (which is slightly superior than soliciting all individuals), and this is because it was not calibrated yet. For the same reason, LAC-MR was also the worst performer in terms of MSE. Calibrated algorithms LAC-MR-NC(N8) and CaSVM showed similar performance in terms of profit. According to [4], the poor performance of CaSVM is, again, due to overfitting. CaDT and CaNB are again in close rivalry. Calibrating LAC-MR using the entropy-minimization method is very profitable, and the corresponding algorithm, LAC-MR-EM is the best performer.

Table 7.4 Comparing algorithms in terms of profit and MSE

Algorithm	Profit	MSE
LAC-MR	$11,097	0.0961
LAC-MR-NC(N8)	$12,442	0.0958
LAC-MR-EM	$14,902	0.0934
CaSVM	$12,969	0.0958
CaNB	$14,682	0.0952
CaDT	$14,190	0.0953

References

1. Boser, B., Guyon, I., Vapnik, V.: A training algorithm for optimal margin classifiers. In: Proceedings of the Annual Conference on Computational Learning Theory (COLT), pp. 144–152. Springer (1992)
2. Cestnik, B.: Estimating probabilities: a crucial task in machine learning. In: Proceedings of the European Conference on Artificial Intelligence (ECAI), pp. 147–149 (1990)
3. Chang, C.-C., Lin, C.-J.: LIBSVM: a library for support vector machines. Available at http://www.csie.ntu.edu.tw/~cjlin/papers/libsvm.pdf (2001)
4. Cohen, I., Goldszmidt, M.: Properties and benefits of calibrated classifiers. In: Proceedings of the European Conference on Principles of Data Mining and Knowledge Discovery (PKDD), pp. 125–136. Springer (2004)
5. Cussens, J.: Bayes and pseudo-bayes estimates of conditional probabilities and their reliability. In: Proceedings of the European Conference on Machine Learning (ECML), pp. 136–152. Springer (1993)
6. DeGroot, M., Fienberg, S.: The comparison and evalution of forecasters. Statistician **32**, 12–22 (1982)
7. Fayyad, U., Irani, K.: Multi interval discretization of continuous-valued attributes for classification learning. In: Proceedings of the International Joint Conference on Artificial Intelligence (IJCAI), pp. 1022–1027 (1993)
8. Platt, J.: Probabilistic outputs for support vector machines and comparison to regularized likelihood methods. Adv. Large Margin Classif. 61–74 (1999)
9. Quinlan, J.: C4.5: Programs for Machine Learning. Morgan Kaufmann, San Francisco (1993)
10. Rissanen, J.: Modeling by shortest data description. Automatica **14**, 465–471 (1978)
11. Zadrozny, B., Elkan, C.: Obtaining calibrated probability estimates from decision trees and naive bayesian classifiers. In: Proceedings of the International Conference on Machine Learning (ICML), pp. 609–616. IEEE Computer Society (2001)

Chapter 8
Self-Training Associative Classification

Abstract The acquisition of training examples usually requires skilled human annotators to manually label the relationship between inputs and outputs. Due to various reasons, annotators may face inputs that are hard to label (Chapelle et al. Semi-supervised learning. MIT Press, Cambridge, 2006). The cost associated with this labeling process thus may render vast amounts of training examples unfeasible. The acquisition of unlabeled inputs (i.e., inputs for which the corresponding output is unknown), on the other hand, is relatively inexpensive. However, it is worthwhile to label at least some inputs, provided that this effort will be then rewarded with an improvement in classification performance. In this chapter demand-driven associative classification will be extended, so that the corresponding algorithm achieves high classification performance even in the case of limited labeling efforts.

Keywords Labeling effort · Unlabeled data · Semi-supervised learning · Reliability · Lack of evidence · Name disambiguation

8.1 Algorithms for Self-Training Associative Classification

Semi-supervised algorithms use few training examples to create additional pseudo-examples, which are used to produce mapping functions. The basic idea adopted by many of these algorithms [3, 9, 10] is to incorporate predictions which are likely to be correct (i.e., highly reliable predictions) into the training data [1, 2]. Thus, the training data is progressively augmented as new unlabeled inputs are processed.

There are several semi-supervised classification algorithms. An excellent survey of the main algorithms is provided by [11].

A. Veloso and W. Meira Jr., *Demand-Driven Associative Classification*,
SpringerBriefs in Computer Science, DOI: 10.1007/978-0-85729-525-5_8,
© Adriano Veloso 2011

In this section we present a self-training demand-driven associative classification algorithm. This algorithm will be referred to as LAC-MR-ST (standing for LAC-MR with self-training). LAC-MR-ST exploits reliable predictions and the lack of enough evidence [4, 5] supporting the known outputs, to include new examples to S.

8.1.1 Reliable Predictions

A reliable prediction, (x_i, c_j) (where $x_i \in T$), is the one for which the corresponding class membership probability, $\hat{p}(c_j|x_i)$, is above a given threshold, Δ_{min}. For appropriate values of Δ_{min}, the chance of $y_i \neq c_j$ (i.e., a misclassification) is low, and thus, these predictions may be exploited for the sake of self-training. In this case, a reliable prediction, (x_i, c_j), is considered as a new example and is added to S. Since rules are extracted on a demand-driven basis, the next input to be processed will possibly take advantage of the recently included (pseudo-)example.

8.1.2 Lack of Evidence

Some problems may contain a very large number of outputs. In such cases, it becomes hard for the annotator to specify all the outputs, and the consequence is that some outputs are never explicitly informed in S. The lack of (enough) decision rules predicting any known output present in S, may be exploited to detect the appearance of a novel/unseen output in T. Specifically, for a given input $x_i \in T$, if the number of rules supporting any known output is smaller than ϕ_{min}, than it is assumed that the input x_i is not related to any output in S. In this case, a new label, c_j, is associated with this possibly new output. The new output, c_j, and the corresponding input, x_i, are considered as a new example, (x_i, c_j), which is included to S.

Algorithm LAC-MR-ST This algorithm exploits reliable predictions and the lack of enough evidence to produce novel training examples, which are stored in N.

Naturally, some predictions are not reliable enough for certain values of Δ_{min}. In these cases, such doubtful predictions are abstained. As new examples are included in N (i.e., the reliable predictions), they may be exploited, hopefully increasing the reliability of the predictions that were previously abstained. To optimize the usage of reliable predictions, inputs are stored in a priority queue, Q, so that inputs having reliable predictions are considered first. The process works as follows. Initially, inputs in T are randomly placed in Q. If the output c_j of the input x_i that is located in the beginning of the queue is reliably predicted, then the x_i is removed from Q and a new example (c_j, x_i) is included into the training data N.

Otherwise, if the prediction is not reliable, the corresponding input is simply placed in the end of the queue and it will be processed again only after processing all other inputs. The process continues performing more reliable predictions first, until no more reliable predictions are possible.

The lack of rules supporting any output in S may be used as evidence indicating the appearance of an output that is not in S. The number of rules that is necessary to consider an output as an already seen one is controlled by a threshold, ϕ_{min}. Specifically, for an input x_i, if the number of rules extracted from S^{x_i}, is smaller than ϕ_{min}, then the output of x_i is considered as an output not in S, and a new label c_j is created to identify such output. Further, this prediction is considered as a new example (x_i, c_j), which is included to N. The basic steps of LAC-MR-ST are shown in Algorithm 16.

Algorithm 16 Including new examples to the original training data.

Require: The training data S, T, σ_{min}, Δ_{min}, and ϕ_{min}
Ensure: N.

1:   $Q \Leftarrow T$
2:   $N \Leftarrow S$
3:   **for each** input $x_i \in Q$ **do**
4:      $\omega \leftarrow$ false
5:      $N^{x_i} \Leftarrow N$ projected according to x_i
6:      $R^{x_i} \Leftarrow$ rules $X \rightarrow c_j$ extracted from N^{x_i}, such that $\pi(X \rightarrow c_j) \geq \sigma_{min} \times |N^{x_i}|$
7:      **if** $|R^{x_i}| < \phi_{min}$ **then**
8:         create a new label c_k
9:         $N \Leftarrow N \cup (x_i{}^c c_k)$
10:        $\omega \leftarrow$ true
11:      **else**
12:         **for each** output c_j **do**
13:            **if** $\hat{p}(c_j|x_i) \geq \Delta_{min}$ **then**
14:               $N \Leftarrow N \cup (x_i{}^c c_j)$
15:               $\omega \leftarrow$ true
16:            **end if**
17:         **end for**
18:      **end if**
19:      **if** $\omega =$ false **then**
20:         place x_i in the end of the queue, Q
21:      **end if**
22:      **if** it is not possible to perform reliable predictions anymore **then return** N
23: **end for**

8.2 Empirical Results

In this section we present experimental results for the evaluation of LAC-MR and LAC-MR-ST.

Setup In all experiments we used ten fold cross-validation, and the final results of each experiment represent the average of the five runs. All the results to be presented were found statistically significant based on a t-test at the 95% confidence level.

Evaluation Criteria Classification performance for the various methods being evaluated is expressed through MicF$_1$ and MacF$_1$.

Baselines We used the k-Way unsupervised Spectral Clustering algorithm as baseline [7].

Parameters For the K-Way Spectral Clustering algorithm we set k to be the correct number of clusters (thus, the performance reported for this algorithm may be considered as an upper-bound of its true performance). For LAC-MR and LAC-MR-ST we set $\sigma_{min} = 0.05$. Particularly for LAC-MR-ST, we investigated its sensitivity to parameters Δ_{min} and ϕ_{min}.

Computational Environment The experiments were performed on a Linux-based PC with a Intel Core 2 Duo 1.83 GHz processor and 2 GB RAM.

8.2.1 Author Name Disambiguation

Citations are an essential component of many current digital libraries. Citation management within digital libraries involves a number of tasks. One task in particular, name disambiguation, has required significant attention from the research community due to its inherent difficulty. Name ambiguity in the context of bibliographic citations occurs when one author can be correctly referred to by multiple name variations (synonyms) or when multiple authors have exactly the same name or share the same name variation (polysems). This problem may occur for a number of reasons, including the lack of standards and common practices, and the decentralized generation of content (e.g., by means of automatic harvesting). Name ambiguity is widespread in many large-scale digital libraries, such as Citeseer, Google Scholar, and DBLP.

Some of the most effective methods seem to be based on the application of supervised machine learning techniques. In this case, the training data consists of examples (x_i, y_i), where x_i is a set of features of a citation, and y_i is a label which identifies the corresponding author. More specifically, such examples are citations for which the correct authorship is known. Although successful cases have been reported [6], some particular challenges associated with name disambiguation in the context of bibliographic citations, prevent the full potential of supervised machine learning techniques:

- The acquisition of training examples requires skilled human annotators to manually label authors in citations. Annotators may face hard-to-label citations with highly ambiguous authors. The cost associated with this labeling process thus may render vast amounts of examples unfeasible. Thus, classification algorithms must be cost-effective, achieving high classification performance even in the case of limited labeling efforts.

Table 8.1 The DBLP and BDBComp collections

DBLP			BDBComp		
Ambiguous group	No. of citations	No. of authors	Ambiguous group	No. of citations	No. of authors
A. Gupta	576	26	A. Oliveira	52	16
A. Kumar	243	14	A. Silva	64	32
C. Chen	798	60	F. Silva	26	20
D. Johnson	368	15	J. Oliveira	48	18
J. Martin	112	16	J. Silva	36	17
J. Robinson	171	12	J. Souza	35	11
J. Smith	921	29	L. Silva	33	18
K. Tanaka	280	10	M. Silva	21	16
M. Brown	153	13	R. Santos	20	16
M. Jones	260	13	R. Silva	28	20
M. Miller	405	12	–	–	–

- It is not reasonable to assume that all possible authors are included in the training data (specially due to the scarce availability of examples). Thus, classification algorithms must be able to detect unseen/unknown authors, for whom no label was previously specified.

We used two collections of bibliographic citations. One was extracted from DBLP (http://dblp.uni-trier.de) and the other was extracted from BDBComp (http://www.lbd.ufmg.br/bdbcomp). Each citation consists of the title of the work, a list of coauthor names, and the title of the publication venue (conference or journal). Pre-processing involved standardizing coauthor names using only the initial letter of the first name along with the full last name, removing punctuation and stop-words of publication and venue titles, stemming publication and venue titles using Porter's algorithm [8], and grouping authors with the same first name initial and the same last name in order to create the ambiguous groups (i.e., groups of citations having different authors with similar names). Table 8.1 shows more detailed information about the collections and their ambiguous groups. Disambiguation is particularly difficult in ambiguous groups such as the C. Chen group, in which the correct author must be selected from 60 possible authors, and in ambiguous groups such as the J. Silva group, in which the majority of authors appears in only one citation.

Analysis In all experiments, we varied the proportion, or fraction of training examples available. For instance, if the fraction of examples available is 0.5, then only half of the examples in the training data was provided to the algorithm. In this case, examples in the training data are randomly selected.

We start our analysis by evaluating the effectiveness of LAC-MR-ST in detecting unseen authors using the BDBComp collection. For each fraction of training examples, we varied ϕ_{min} from 1 to 6. The results are shown in Fig. 8.1, where each curve is associated with a different fraction of training examples. For the BDBComp collection, the fraction of unseen authors that are detected increases

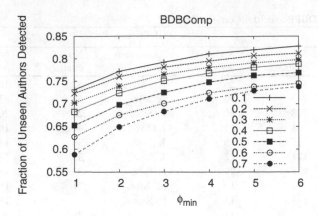

Fig. 8.1 Sensitivity to ϕ_{min}

Fig. 8.2 Sensitivity to Δ_{min}

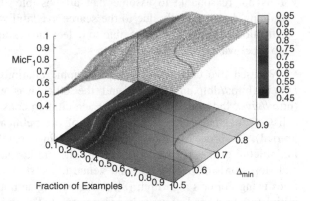

with ϕ_{min}. This is expected, since the amount of evidence that is required to recognize an author as already seen one, increases for higher values of ϕ_{min}. Further, it becomes more difficult to detect an unseen author when the fraction of training examples increases. This is because, in such cases, (1) more authors are seen (i.e., there are more examples), and (2) there is an increase in the amount of available evidence supporting already seen authors.

We evaluate the effectiveness of LAC-MR-ST in incorporating new training examples using the DBLP collection. For each fraction of training examples, we varied Δ_{min} from 0.5 to 0.9. The results are shown in Fig. 8.2. As it can be seen, the performance of LAC-MR-ST decreases when Δ_{min} is set too high (i.e., $\Delta_{min} > 0.75$). Further, the performance also decreases when Δ_{min} is set too low (i.e., $\Delta_{min} < 0.65$). On one hand, when lower values of Δ_{min} are applied, several citations in the test set, which are associated with wrong predictions, are included in the training data, hurting performance. On the other hand, when higher values of Δ_{min} are applied, only few citations in the test set are included in the training data.

Table 8.2 MicF$_1$ numbers for DBLP collection

Ambiguous group	LAC-MR-ST	K-way SC
A. Gupta	0.453	**0.546**
A. Kumar	**0.555**	0.505
C. Chen	0.365	**0.607**
D. Johnson	**0.710**	0.561
J. Martin	0.786	**0.939**
J. Robinson	**0.662**	0.693
J. Smith	0.444	**0.500**
K. Tanaka	0.554	**0.626**
M. Brown	0.680	**0.759**
M. Jones	**0.504**	0.628
M. Miller	**0.699**	0.479
Average	**0.583**	**0.622**

For the DBLP collection, LAC-MR-ST achieves the best performance when Δ_{min} is between 0.65 and 0.75 (specially when few training examples are available).

We now evaluate how the self-training ability of LAC-MR-ST improves its performance when compared with LAC-MR. Figure 8.3 shows some of the results. The value associated with each point in each graph is obtained by applying a different combination of Δ_{min} and ϕ_{min}, for different fractions of training examples. For the DBLP collection, gains ranging from 18.4 to 53.8% are observed when few training examples are available. The improvement decreases as more examples are available, since in this case (1) more authors are seen and (2) additional examples that are included in the training data do not impact so much the final performance. Interestingly, LAC-MR-ST achieves good performance even when not a single example is available for training. This is possible because, in this case, citations authored by unseen authors are included in the training data, and used as training examples. These gains highlight the advantages of self-training.

Improvements obtained using the BDBComp collection are more impressive. This collection contains several authors that appear in only one citation. LAC-MR is not useful in such scenarios. (i.e., if this citation appears in the test set, then the training data contains no evidence supporting the correct author). LAC-MR-ST, on the other hand, is highly effective in such cases, being able to detect unseen authors, and to make use of this information to enhance the training data with additional examples. As a result, improvements provided by LAC-MR-ST range from 241.6 to 407.1%. Thus, LAC-MR-ST is not only able to reduce labeling efforts (as shown in the experiments with the DBLP collection), but it is also able to detect novel and important information (i.e., unseen authors), being highly practical and effective in a variety of scenarios.

In the next experiment, we used the DBLP collection to perform a comparison between LAC-MR-ST ($\Delta_{min} = 0.7$, $\phi_{min} = 4$), and the k-way Spectral Clustering algorithm [7], when no training example is available. We adopted the evaluation methodology proposed in [7], so that we can directly compare the performance of both algorithms. In this case, a confusion matrix is used to assess MicF$_1$ numbers.

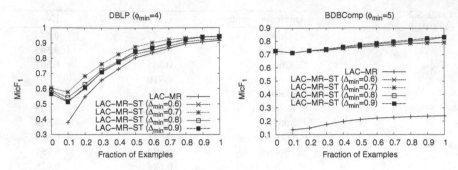

Fig. 8.3 MicF$_1$ values for different Δ_{min} and ϕ_{min}

A different confusion matrix is associated with each ambiguous group, and the final performance is represented by the accuracy averaged over all groups.

Table 8.2 shows the results. Best results, including statistical ties, are highlighted in bold. As it can be seen, both algorithms provide results that are statistically tied on almost all ambiguous groups. The K-way spectral clustering algorithm obtained superior performance on three ambiguous groups, while LAC-MR-ST was superior in one ambiguous group. It is important to notice that the k-way spectral clustering algorithm takes as input the correct number of clusters to be generated, that is, if there are k authors in a group, then this group is clustered into exactly k clusters [7]. This is clearly unrealistic in an actual or practical scenario, but provides something closer to an upper-bound for an unsupervised algorithm that has privileged information. LAC-MR-ST, on the other hand, does not use this information, and works by detecting unseen authors, and incrementally adding new examples to the training data. Other point worth mentioning is that, as shown in Fig. 8.3, with small labeling efforts, the performance of LAC-MR-ST is improved (greatly outperforming the un algorithm), demonstrating that LAC-MR-ST is cost-effective (the only exception is when only very few examples are available, because in this case it seems that LAC-MR-ST has some difficulties in detecting novel authors, hurting disambiguation performance).

References

1. Bennett, K., Demiriz, A.: Semi-supervised support vector machines. In: Proceedings of the Annual Conference on Neural Information Processing Systems (NIPS), pp. 368–374 (1998)
2. Blum, A., Chawla, S.: Learning from labeled and unlabeled data using graph mincuts. In: Proceedings of the International Conference on Machine Learning (ICML), pp. 19–26. IEEE Computer Society (2001)
3. Blum, A., Mitchell,T.: Combining labeled and unlabeled sata with co-training. In: Proceedings of the Annual Conference on Computational Learning Theory (COLT), pp. 92–100. Springer (1998)
4. Ferreira, A., Gonçalves, M., Almeida, J., Laender, A., Veloso, A.: Sygar—a synthetic data generator for evaluating name disambiguation methods. In: Proceedings of the European

Conference on Research and Advanced Technology for Digital Libraries (ECDL), pp. 437–441. Springer (2009)

5. Ferreira, A., Veloso, A., Gonçalves, M., Laender, A.: Effective self-training author name disambiguation in scholarly digital libraries. In: Proceedings of the Joint Conference on Digital Libraries (JCDL), pp. 39–48 (2010)

6. Han, H., Giles, C.L., Zha, H., Li, C., Tsioutsiouliklis, K.: Two supervised learning approaches for name disambiguation in author citations. In: Proceedings of the Joint Conference on Digital Libraries (JCDL), pp. 296–305. ACM Press (2004)

7. Han, H., Zha, H., Giles, C.L.: Name disambiguation in author citations using a k-way spectral clustering method. In: Proceedings of the Joint Conference on Digital Libraries (JCDL), pp. 334–343. ACM Press (2005)

8. Porter, M.F.: An algorithm for suffix stripping. Program **14**(3), 130–137 (1980)

9. Tresp, V.: A bayesian committee machine. Neural Comput. **12**(11), 2719–2741 (2000)

10. Vapnik, V.: Statistical Learning Theory. Wiley, New York (1998)

11. Zhu, X.: Semi-supervised learning literature survey. Technical report, University of Winsconsin, Computer Sciences, TR 150 (2008)

Chapter 9
Ordinal Regression and Ranking

Abstract Accurate ordering or ranking over instances is of paramount importance for several applications (Faria et al. Learning to rank for content-based image retrieval. In: Proceedings of the Multimedia Information Retrieval Conference, pp. 285–294, 2010; Veloso et al. Learning to rank at query-time using association rules. In: Proceedings of the Conference on Research and Development in Information Retrieval (SIGIR), pp. 267–274, 2008; Veloso et al. J Inf Data Manag 1(3): 567–582, 2010; Veloso and Meira, Efficient on-demand opinion mining. In: Proceedings of the Brazilian Symposium on Databases (SBBD), pp. 332–346, 2007; Veloso et al. Automatic moderation of comments in a large on-line journalistic environment. In: Proceedings of the International AAAI Conference on Weblogs and Social Media (ICWSM), pp. 234–237, AAAI, 2007).One clear application is Information Retrieval, where documents retrieved by search engines must be ranked according to the corresponding relevance to the query (Trotman, Inf Ret 8(3):359–381, 2005).Many features may affect the relevance of such documents, and, thus, it is difficult to adapt ranking functions manually. Recently, a body of empirical evidence has emerged suggesting that methods that automatically learn ranking functions offer substantial improvements in enough situations to be regarded as a relevant advance for applications that depend on ranking. Hence, learning ranking functions has attracted significant interest from the machine learning community. In the context of Information Retrieval, the conventional approach to this learning task is to assume the availability of examples (i.e., a training data, S, which typically consists of document features and the corresponding relevance to specific queries), from which a learning function can be learned. When a new query is given, the documents associated with this query are ranked according to the learned function (i.e., this function gives a score to a document indicating its relevance with regard to the query). In this chapter we present ranking algorithms based on demand-driven associative classification.

A. Veloso and W. Meira Jr., *Demand-Driven Associative Classification*,   97
SpringerBriefs in Computer Science, DOI: 10.1007/978-0-85729-525-5_9,
© Adriano Veloso 2011

Keywords Ranking function · Relevance · Likelihood of membership · Search engine · Information retrieval · Learning to rank

9.1 Algorithms for Ordinal Regression and Ranking

In this section, we present an algorithm which learns ranking functions for Information Retrieval, based on demand-driven associative classification. This algorithm will be referred to as LAC-MR-OR (standing for LAC-MR for ordinal regression).

Algorithm LAC-MR-OR Extending demand-driven associative classification algorithms to sort inputs is rather simple [7]. The first step is to extract from S, decision rules associating inputs to outputs. In the context of Information Retrieval, inputs are document features and outputs are relevance levels. These rules are extracted on a demand-driven basis, as described in Sect. 4.1. The next step is to calculate, using Eq. 3.5, the likelihood of each relevance level (i.e., c_j) for each document (i.e., x_i).

Finally, the ranking position of input x_i can be estimated by a linear combination of the likelihoods associated with each output (or each relevance level), as shown in Eq. 9.1. Higher values of $rank(x_i)$ indicate that input x_i should be placed in first positions of the rank. Basic steps of LAC-MR-OR are shown in Algorithm 17.

$$rank(x_i) = \sum_{j=0}^{p} \left(c_j \times \hat{p}(c_j|x_i) \right) \tag{9.1}$$

Algorithm 17 Producing ranking scores using LAC-MR-OR.

Require: The training data S, input $x_i \in T$, σ_{min}
Ensure: $rank(x_i)$

1: $S^{x_i} \Leftarrow S$ projected according to x_i
2: $R^{x_i} \Leftarrow$ rules $X \rightarrow c_j$ extracted from S^{x_i}, such that $\pi(X \rightarrow c_j) \geq \sigma_{min} \times |S^{x_i}|$
3: calculate membership probabilities $\hat{p}(c_j|x_i)$ for each output c_j
4: return $\sum_{j=0}^{p} \left(c_j \times \hat{p}(c_j|x_i) \right)$

9.2 Empirical Results

In this section, we empirically analyze the proposed algorithm, LAC-MR-OR. We first present the collections employed in the evaluation, and then we discuss the effectiveness of LAC-MR-OR in these collections.

Setup In all experiments we used fivefold cross-validation, and the final results of each experiment represent the average of the five runs. All the results to be presented were found statistically significant based on a t-test at the 95% confidence level.

Evaluation Criteria Ranking performance is evaluated using NDCG@k, P@k, and MAP measures. Basically, these measures express the ability to place documents with high relevance in the first positions of the ranking. If the set of relevant documents for a query $q_j \in Q$ is $\{x_1, \ldots, x_{m_j}\}$, and D_j is the set of ranked documents associated with query q_j, then P@k is defined in Eq. 9.2 (where $r(x_i)$ is the true relevance of document $x_i \in D_j$). Precision values are averaged over all queries.

$$P@k(D_j) = \sum_{i=1}^{k} \frac{r(x_i)}{k} \qquad (9.2)$$

For a single query, average precision (AP) is the average of the precision values obtained for the set of top k documents existing after each relevant document is retrieved, as given in Eq. 9.3. MAP is obtained by averaging AP values over all queries, as shown in Eq. 9.3.

$$MAP(Q) = \frac{1}{|Q|} \sum_{j=1}^{|Q|} \frac{1}{m_j} \sum_{k=1}^{m_j} P@k(D_j) \qquad (9.3)$$

Finally, NDCG@k is defined in Eq. 9.4, where Z_k is a normalization factor calculated to make it so that the NDCG value of a perfect ranking is 1. NDCG values are averaged over all queries.

$$NDCG@k(D_j) = Z_k \sum_{i=1}^{k} \frac{2^{rank(x_i)} - 1}{\log(1 + i)} \qquad (9.4)$$

Computational Environment The experiments were performed on a Linux-based PC with a Intel Pentium III 1.0 GHz processor and 1 GB RAM.

Baselines and Parameters Our evaluation is based on a comparison against state-of-the-art learning to rank algorithms such as R-SVM [9], FRank [6], R-Boost [3], SVMMAP [4], AdaRank [8], and ListNet [1].

9.2.1 Learning to Rank

LETOR [5] is a benchmark for research on learning to rank, released by Microsoft Research Asia.[1] It makes available seven subsets (OHSUMED, TD2003, TD2004, HP2003, HP2004, NP2003 and NP2004). Each subset contains a set of queries, document features, and the corresponding relevance judgments. Features cover a wide range of properties, such as term frequency, BM25, PageRank, HITS, etc. Documents are given as inputs (i.e., x_i), and their relevance levels are the corresponding outputs (i.e., c_j). The goal is to place relevant documents in the first

---

[1] LETOR Web page: http://research.microsoft.com/users/LETOR/

Table 9.1 MAP numbers for OHSUMED subset

Trial	LAC-MR-OR	R-SVM	R-Boost	FRank	ListNet	AdaRank	SVMMAP
1	**0.352**	0.304	0.332	0.333	**0.346**	0.344	0.342
2	**0.463**	0.447	0.445	0.438	0.450	0.446	0.454
3	0.460	**0.465**	0.456	0.456	0.461	**0.469**	**0.462**
4	**0.521**	0.499	0.508	0.513	0.511	0.514	0.518
5	**0.482**	0.453	0.464	**0.481**	0.461	0.471	0.450
Avg	**0.456**	0.433	0.441	0.444	0.446	0.449	0.445

Table 9.2 MAP numbers for TD2003 subset

Trial	LAC-MR-OR	R-SVM	R-Boost	FRank	ListNet	AdaRank	SVMMAP
1	0.169	0.164	0.110	0.113	**0.192**	0.153	0.172
2	0.293	0.258	0.291	0.297	**0.325**	0.251	0.237
3	0.365	**0.408**	0.251	0.155	0.381	0.290	0.342
4	**0.394**	0.236	0.262	0.212	0.275	0.322	0.276
5	0.219	**0.249**	0.222	0.238	0.202	0.125	0.196
Avg	**0.288**	0.263	0.227	0.203	0.275	0.228	0.244

Table 9.3 MAP numbers for TD2004 subset

Trial	LAC-MR-OR	R-SVM	R-Boost	FRank	ListNet	AdaRank	SVMMAP
1	0.213	0.211	**0.247**	0.226	0.225	0.173	0.185
2	**0.276**	0.209	**0.281**	0.203	0.215	0.248	0.192
3	**0.285**	0.206	0.241	0.218	0.223	0.229	0.201
4	0.267	0.218	0.238	**0.285**	0.223	0.194	0.211
5	0.276	0.274	**0.299**	0.262	0.229	0.250	0.235
Avg	**0.263**	0.224	**0.261**	0.239	0.223	0.219	0.205

Table 9.4 MAP numbers for NP2003 subset

Trial	LAC-MR-OR	R-SVM	R-Boost	FRank	ListNet	AdaRank	SVMMAP
1	**0.695**	0.625	0.685	0.591	0.593	0.621	0.623
2	**0.676**	0.662	0.666	0.645	0.648	0.620	0.640
3	0.670	0.695	0.711	0.673	**0.751**	0.660	0.714
4	0.751	**0.761**	0.733	**0.769**	0.724	0.702	0.736
5	0.748	0.735	0.743	0.642	0.732	**0.789**	0.721
Avg	**0.708**	0.695	**0.707**	0.664	0.689	0.678	0.687

positions of the ranking. Pre-processing involved only the discretization [2] of attribute-values in S.

Tables 9.1, 9.2, 9.3, 9.4, 9.5, 9.6, and 9.7 show MAP numbers for the seven subsets. Best results, including statistical ties, are shown in bold. The result for each trial is obtained by averaging partial results obtained from each query in the

Table 9.5 MAP numbers for NP2004 subset

Trial	LAC-MR-OR	R-SVM	R-Boost	FRank	ListNet	AdaRank	SVMMAP
1	0.592	0.535	0.550	0.599	0.550	**0.700**	0.574
2	0.648	0.608	0.559	0.629	0.659	0.594	**0.669**
3	**0.870**	0.756	0.609	0.731	0.739	0.607	0.767
4	0.611	0.694	0.531	0.485	**0.728**	0.600	0.599
5	0.650	**0.701**	0.570	0.560	0.684	0.608	**0.701**
Avg	**0.675**	0.659	0.564	0.601	**0.672**	0.622	0.662

Table 9.6 MAP numbers for HP2003 subset

Trial	LAC-MR-OR	R-SVM	R-Boost	FRank	ListNet	AdaRank	SVMMAP
1	**0.717**	0.684	0.634	0.674	0.728	0.715	**0.729**
2	0.808	0.796	0.813	0.804	**0.852**	**0.855**	0.775
3	0.737	0.783	0.781	0.737	**0.821**	0.801	0.785
4	**0.762**	**0.763**	0.745	0.684	**0.772**	0.752	0.719
5	**0.755**	0.679	0.692	0.648	0.657	0.732	0.579
Avg	0.756	0.741	0.733	0.709	0.766	**0.771**	0.717

Table 9.7 MAP numbers for HP2004 subset

Trial	LAC-MR-OR	R-SVM	R-Boost	FRank	ListNet	AdaRank	SVMMAP
1	0.666	0.664	0.634	0.674	**0.728**	0.715	**0.717**
2	0.756	0.680	0.813	0.804	**0.852**	**0.855**	**0.845**
3	0.806	0.742	0.781	0.737	**0.821**	0.801	0.780
4	0.635	0.715	0.745	0.684	**0.772**	0.752	0.760
5	0.627	0.536	0.692	0.648	0.657	**0.732**	0.608
Avg	0.696	0.667	0.733	0.709	**0.766**	**0.771**	0.742

trial. The final result is obtained by averaging the five trials. We conducted two sets of significance tests (t-test) on each subset. The first set of significance tests was carried on the average of the results for each query. The second set of significance tests was carried on the average of the five trials.

In five, out of seven subsets, LAC-MR-OR was the best overall performer, demonstrating the effectiveness of demand-driven associative classification. In most of the subsets, LAC-MR-OR achieved superior ranking performance when compared to the best baseline. The only exceptions occurred in HP2003 and HP2004 subsets, where AdaRank was the best performer. Still, LAC-MR-OR obtained a ranking performance which is much better than the performance obtained by the worst baselines. Gains provided by LAC-MR-OR range from 6.6% (relative to FRank in NP2003) to 42% (relative to FRank in TD2003).

We also evaluated LAC-MR-OR in terms of precision and NDCG. Figure 9.1 shows precision numbers obtained from the execution of LAC-MR-OR. LAC-MR-OR improved the precision at the first positions (it was always the best performer

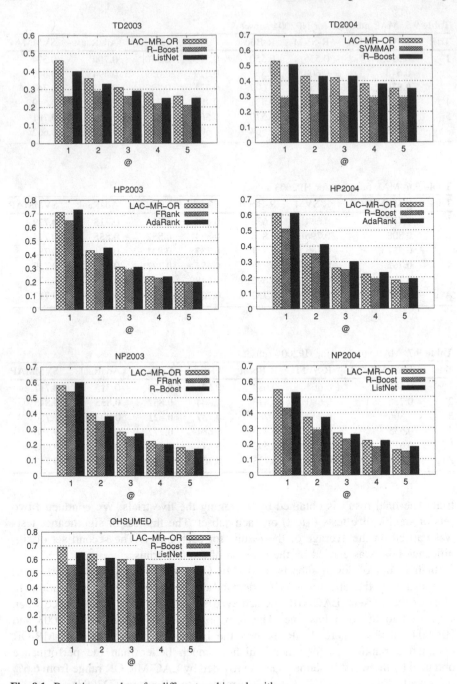

Fig. 9.1 Precision numbers for different ranking algorithms

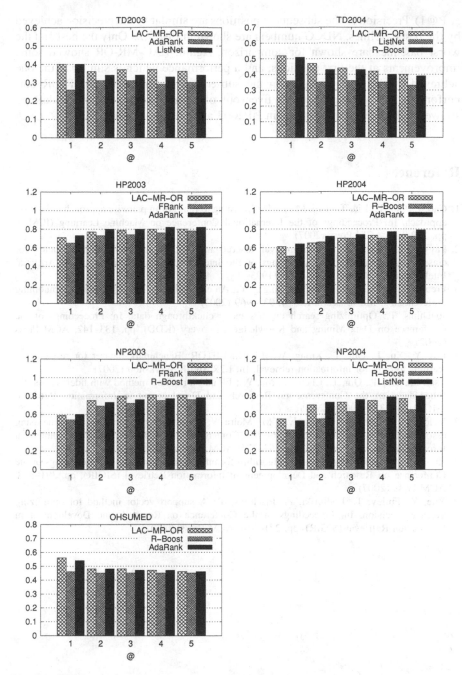

Fig. 9.2 NDCG Numbers for different ranking algorithms

at P@1). Precision in the subsequent positions are similar to the precision achieved by the best baselines. NDCG numbers are shown in Fig. 9.2. Only the best and the worst baselines are shown for comparison. Again, LAC-MR-OR showed some improvements at the first positions, and a performance which is similar to the one achieved by the best baselines in the subsequent positions. LAC-MR-OR outperformed the best baselines in five (out of seven) subsets. Again, AdaRank showed to be the best performer in HP2003 and HP2004 subsets.

References

1. Cao, Z., Qin, T., Liu, T., Tsai, M., Li, H.: Learning to rank: from pairwise approach to listwise approach. In: Proceedings of the International Conference on Machine Learning (ICML), pp. 129–136. ACM Press (2007)
2. Fayyad, U., Irani, K.: Multi interval discretization of continuous-valued attributes for classification learning. In: Proceedings of the International Joint Conference on Artificial Intelligence (IJCAI), pp. 1022–1027 (1993)
3. Freund, Y., Iyer, R., Schapire, R., Singer, Y.: An efficient boosting algorithm for combining preferences. J. Mach. Learn. Res. **4**, 933–969 (2003)
4. Joachims, T.: Optimizing search engines using clickthrough data. In: Proceedings of the Conference on Data Mining and Knowledge Discovery (KDD), pp. 133–142. ACM Press (2002)
5. Liu, Y., Xu, J., Qin, T., Xiong, W., Li, H.: LETOR: Benchmark dataset for research on learning to rank for information retrieval. In: L2R SIGIR Workshop (2007)
6. Tsai, M., Liu, T., Qin, T., Chen, H., Ma, W.: FRank: a ranking method with fidelity loss. In: Proceedings of the Conference on Research and Development in Information Retrieval (SIGIR), pp. 383–390. ACM Press (2007)
7. Veloso, A., Almeida, H., Gonçalves, M., Meira W. Jr.: Learning to rank at query-time using association rules. In: Proceedings of the Conference on Research and Development in Information Retrieval (SIGIR), pp. 267–274. ACM Press (2008)
8. Xu, J., Li, H.: Adarank: a boosting algorithm for information retrieval. In: Proceedings of the Conference on Research and Development in Information Retrieval (SIGIR), pp. 391–398. ACM Press (2007)
9. Yue, Y., Finley, T., Radlinski, F., Joachims, T.: A support vector method for optimizing average precision. In: Proceedings of the Conference on Research and Development in Information Retrieval (SIGIR), pp. 271–278. ACM Press(2007)

Part IV
Conclusions and Future Work

Chapter 10
Conclusions

Abstract In this chapter we summarize the research contributions of this work and point out limitations and problems that remained open.

Keywords Limitations · Classification time · Minimum support · Generalization bounds · Parallelism · Condensed representation

10.1 Summary of Results

The basic problem discussed in this book is known as classification, in which it is given a set of inputs and outputs, that are somehow related. The goal is to produce a mapping function that approximates this relationship, so that this function is used to predict the outputs for unknown inputs. We examined this problem from a function approximation perspective and proposed several classification algorithms. The first algorithm was shown to be PAC-efficient, and continuous improvements have led to the other algorithms introduced in this book. Such improvements resulted in demand-driven associative classification algorithms. We have shown that these algorithms produce functions that provide high classification performance in several real-world problems. The key insight that has led to these algorithms is that solving sub-problems may be much easier than directly solving the entire problem. Furthermore, we have shown that, frequently, each sub-problem demands approximation strategies that are very different from the strategy adopted when the entire problem is solved at once. Thus, producing approximation functions on a demand-driven basis may provide finer-grained results that are not achievable when the original problem is not broken into sub-problems. We successfully extended demand-driven associative classification algorithms to solve a number of problems that are related to the original classification problem, including cost-sensitive and cautious classification, multi-label classification,

A. Veloso and W. Meira Jr., *Demand-Driven Associative Classification*,
SpringerBriefs in Computer Science, DOI: 10.1007/978-0-85729-525-5_10,
© Adriano Veloso 2011

multi-metric classification, classification with limited labeling efforts, and ordinal regression.

10.2 Limitations

The proposed demand-driven associative classification algorithms have some limitations when compared to other classification techniques. These include:

1. Off-line discretization: the proposed algorithms are not able to process continuous attributes directly. First, these attributes must be discretized. We noticed that the classification performance of the proposed algorithms depends on how effectively attributes are discretized. Supervised discretization techniques, such as [7], have shown to be effective in many cases. However, these techniques do not exploit the correlation among attribute-values, and this information may be lost. There are also discretization techniques that preserve the correlation between different attributes while producing the intervals [10], but these are not supervised, and consequently, they may not consider important information about the correlation between inputs and outputs while producing intervals. Apart from this, discretization is a pre-processing step that several other classification algorithms (SVMs, KNN, many decision trees, etc.) do not need to perform.
2. Classification time: the proposed algorithms perform almost all the computation at classification time. While this strategy enables a great decrease in the total execution time, the classification time inevitably increases. We have shown that caching is extremely effective in keeping classification time low. However, this still may cause problems, for instance, in real-time applications, where adverse situations may take place, possibly increasing classification time.

10.3 Open Problems

Several interesting problems remained open. These problems include:

1. The choice of σ_{min}: we have shown how to adapt cut-off values according to each sub-problem. However, the choice of the σ_{min} threshold that leads to the best classification performance is still an open problem. Interestingly, different values of σ_{min} induce nested classes of functions (i.e., functions derived from rules with support higher than σ_{min}). Thus, the same strategy we used to select the proper complexity of a function (i.e., Empirical/Structural risk minimization), can also be used to set σ_{min}.
2. Generalization bounds for associative classification algorithms: we have used some general bounds derived from the stability or from the VC-dimension of a

function. Specific bounds for associative classification algorithms can be derived using inputs in the test set. Specifically, we can confront how different features are associated with each other, in the training data and in the test set. The discrepancy observed between associations in the training data and associations in the test set may provide a powerful tool that may be used to bound generalization.

3. Semi-supervised associative classification algorithms: calibrated probabilities seems to be valuable in applications where few training examples are provided to the classification algorithm. During the classification of an input in the test set, if the probability associated with an output is substantially larger than the probabilities associated with the other outputs, then, in practice, the chance of misclassification is very low. In such cases, the input, along with the predicted output, can be incorporated to the training data (with low risk), increasing the number of examples and potentially improving classification performance. This is because, usually, features in inputs in the training data are redundantly sufficient to describe the examples, and thus associations between features of inputs in the training data and in the test set may be exploited.

4. Parallel associative classification algorithms: the use of large amounts of training data can enable the achievement of highly accurate mapping functions [6]. Thus, its is necessary the development of high-performance scalable classification algorithms, which are able to process large amounts of training data efficiently (which eventually may not fit in main memory [16]). The algorithms introduced in this book may be greatly improved by parallel processing. First, each input in the test set induces a sub-problem, and each sub-problem may be processed independently from each other. This fact enables the use of a simple "bag of tasks" strategy, in which each sub-problem corresponds to a task. To make the process asynchronous, each processor will use a separate cache for storing its own decision rules. Load balancing is optimized, since processors will become idle only if no more tasks are available (i.e., there is no more inputs in the test set). Other degrees of parallelism may be further explored. For instance, the process of extracting rules from the training data can be efficiently parallelized following the strategies proposed in [12–14].

5. Excessively large rule sets: it is challenging to store, retrieve and sort a large number of rules efficiently for classification. The large number of rule also compromises model interpretability. Instead of using the entire set of useful rules, we could use a condensed representation. Among these condensed representations, there are rule sets derived from maximal feature-sets [4, 8], closed feature-sets [2, 15], and non-derivable feature-sets [5, 11]. There are associative classification algorithms that employ closed feature-sets in order to eliminate redundant rules [1], and we intend to evaluate the impact and trade-offs of using other representations such as maximal, non-derivable, δ-free [3] and negative disjunction-free [9] features-sets.

References

1. Baralis, E., Chiusano, S.: Essential classification rule sets. Trans. Database Syst. **29**(4), 635–674 (2004)
2. Bastide, Y., Pasquier, N., Taouil, R., Stumme, G., Lakhal, L.: Mining minimal non-redundant association rules using frequent closed itemsets. In: International Conference on Computational Logic, pp. 972–986 (2000)
3. Boulicaut, J., Bykowski, A., Rigotti, C.: Free-sets: A condensed representation of boolean data for the approximation of frequency queries. Data Min. Knowl. Discov. **7**(1), 5–22 (2003)
4. Burdick, D., Calimlim, M., Flannick, J., Gehrke, J., Yiu, T.: Mafia: A maximal frequent itemset algorithm. IEEE Trans. Knowl. Data Eng. **17**(11), 1490–1504 (2005)
5. Calders, T., Goethals, B.: Non-derivable itemset mining. Data Min. Knowl. Discov. **14**(1), 171–206 (2007)
6. Chan, P., Stolfo, S.: Experiments on multistrategy learning by meta-learning. In: Proceedings of the Conference on Information and Knowledge Management (CIKM), pp. 314–323. ACM Press (1993)
7. Fayyad, U., Irani, K.: Multi interval discretization of continuous-valued attributes for classification learning. In: Proceedings of the International Joint Conference on Artificial Intelligence (IJCAI), pp. 1022–1027 (1993)
8. Gouda, K., Zaki, M.: Efficiently mining maximal frequent itemsets. In: Proceedings of the International Conference on Data Mining (ICDM). IEEE Computer Society, (2001)
9. Kryszkiewicz, M.:Generalized disjunction-free representation of frequent patterns with negation. J. Exp. Theor. Artif. Intell. **17**(1–2), 63–82 (2005)
10. Mehta, S., Parthasarathy, S., Yang, H.: Toward unsupervised correlation preserving discretization. Trans. Knowl. Data Eng. **17**(9), 1174–1185 (2005)
11. Muhonen, J., Toivonen, H. Closed non-derivable itemsets. In: Proceedings of the European Conference on Principles of Data Mining and Knowledge Discovery (PKDD), pp. 601–608 (2006)
12. Otey, M., Parthasarathy, S., Wang, C., Veloso, A.., Meira, Jr W.: Parallel and distributed methods for incremental frequent itemset mining. Trans. Syst. Man Cybern., Part B, **34**(6), 2439–2450 (2004)
13. Otey, M.E., Wang, C., Parthasarathy, S., Veloso, A., Meira, W. Jr. : Mining frequent itemsets in distributed and dynamic databases. In: Procceedings of the International Conference on Data Mining (ICDM), IEEE Computer Society, pp. 617–620 (2003)
14. Veloso, A.., Meira, Jr. W., Ferreira, R., Guedes, D., Parthasarathy, S.: Asynchronous and anticipatory filter-stream based parallel algorithm for frequent itemset mining. In: Proceedings of the European Conference on Principles of Data Mining and Knowledge Discovery (PKDD), pp. 422–433. Springer (2004)
15. Zaki, M.: Mining non-redundant association rules. Data Min. Knowl. Discov. **9**(3), 223–248 (2004)
16. Zaki, M., Ho, C., Agrawal, R.: Parallel classification for data mining on shared-memory multiprocessors. In: Proceedings of the International Conference on Data Engineering (ICDE), IEEE Computer Society, pp. 198–205 (1999)

Index